1996

ADVANCES IN
DETAILED REACTION MECHANISMS
Radical, Single Electron Transfer, and
Concerted Reactions

Volume 1 • 1991

ADVANCES IN
DETAILED REACTION MECHANISMS
Radical, Single Electron Transfer, and Concerted Reactions

A Research Annual

Editor: JAMES M. COXON
University of Canterbury
Christchurch, New Zealand

VOLUME 1 • 1991

 JAI PRESS INC.

Greenwich, Connecticut   *London, England*

CONTENTS

LIST OF CONTRIBUTORS

Michael J. Davies

Department of Chemistry
University of York
Heslington, York, United Kingdom

William R. Dolbier, Jr.

Department of Chemistry
University of Florida
Gainesville, Florida

Christopher J. Easton

Organic Chemistry Department
University of Adelaide
Adelaide, South Australia

Bruce C. Gilbert

Department of Chemistry
University of York
Heslington, York, United Kingdom

Martin Newcomb

Department of Chemistry
Wayne State University
Detroit, Michigan

INTRODUCTION TO THE SERIES:
AN EDITOR'S FOREWORD

The field of organic chemistry has developed dramatically during the past forty years. Thus it appears to be an opportune time to publish a series of essays on various relevant themes written by workers who are active in the discipline. This collection includes many of the important areas of current research interest. To cover such a broad area a very substantial effort was needed, as was the cooperation of a large number of colleagues and friends who have agreed to act as series editors. I have been gratified by the favorable response of research workers in the field to the invitation to contribute chapters in their own specialties. Each contributor has written a critical, lively, and up-to-date description of his field of interest and competence, so that the chapters are not merely literature surveys. It is hoped that this new and continuing series will prove valuable to active researchers, and that many ideas will be generated for future theoretical and experimental research. The wide coverage of material should be of interest to graduate students, postdoctoral fellows, and those teaching specialized topics to graduate students.

Department of Chemistry   Albert Padwa
Emory University   *Consulting Editor*
Atlanta, Georgia

PREFACE

The study of *detailed reaction mechanisms*, of how and why molecular change occurs, forms the basis of this series. It is intended to highlight selected developments that have led to advances in the understanding and control of nature. This first volume details reactions in which *radical, single-electron transfer, and concerted pathways* have been identified in chemical change.

Four critical reviews report the unique properties of reactions that enable them to be harnessed in simple yet elegant approaches to synthetic strategies: nucleophilic substitution, in which single-electron transfer is important; the plethora of reaction possibilities for radicals in aqueous solution; the study of the radicals of the amino acids; and cycloaddition, in which reaction in concert competes with biradical pathways.

These chapters represent much of each author's life's work to date and illustrate ingenuity, resourcefulness, and imagination in probing nature, which has often yielded its secrets only reluctantly. With a basis of sound mechanistic understanding, selective control of synthesis becomes possible and the return rewards the labor. The record of achievement in the quest for "why" and "how" testifies to the human spirit and curiosity, promising hope in the continuing search for understanding.

Future volumes will continue to cover the variety of detailed reaction mechanisms and molecular change.

James M. Coxon
Series Editor

RADICAL KINETICS AND MECHANISTIC PROBE STUDIES

Martin Newcomb

Advances in Detailed Reaction Mechanisms
Volume 1, pages 1–33
Copyright © 1991 JAI Press Inc.
All rights of reproduction in any form reserved.
ISBN: 1-55938-164-7

1. INTRODUCTION

It is an unquestioned point among chemists in any subfield that electron transfer processes comprise a major group of chemical reactions. In the context of organic reaction mechanisms, the prevalence of electron transfer (ET) pathways has been a subject of substantial recent interest despite the fact that the field is perceived to be relatively well developed. This review is concerned with ET processes that might occur when organic substrates react with "anionic" reagents that generally are regarded as strong bases or nucleophiles. Historically, these reactions have been described in terms of polar reaction mechanisms wherein bond-breaking and bond-forming steps are "two-electron" processes. In the latter half of the 1970s and throughout the 1980s, such mechanistic conclusions were questioned for a substantial number of reactions. A school of thought developed that held that many reactions previously thought to proceed by polar pathways actually occurred by sequences of reactions that involved electron transfer processes and radical intermediates. The pathways were broadly labeled as "single-electron transfer" (SET) mechanisms. The bolder advocates of SET reaction pathways argued that even some deeply entrenched polar reaction mechanisms like bimolecular nucleophilic substitutions (S_N2 reactions) and nucleophilic additions to carbonyl groups actually proceeded via pathways that involved initial ET from the reagent to the organic substrate and formation of radical intermediates.

2. MECHANISTIC PROBES AND SET REACTION PATHWAYS

A prominent method of investigation of potential ET pathways in reactions of alkyl halides with nucleophiles involves the use of the so-called "mechanistic probes." The probes are alkyl halides that are precursors to radicals that can undergo a fast rearrangement. The essence of the study is to allow the probe halide to react with a nucleophile, and then to analyze the products to determine if rearranged substitution products were formed. In principle, any radical rearrangement could be used as a test, with the only limitations concerning the rate of the rearrangement and a requirement that the rearrangement can be shown to be unique to the radical. A variety of radical reactions can be envisioned for use in probe studies, including configurational isomerizations, ring openings, ring closures, and group migrations. In practice, radical cyclizations onto a δ,ε-unsaturated site, like the 5-exo cyclization of the 5-hexenyl radical shown in Figure 1, have been most popular. These cyclizations are relatively fast ($k \geq 1 \times 10^5$ s^{-1} at 25°C) and essentially unique to the radical because the corresponding cation cyclizes mainly in a 6-endo fashion and the corresponding anion cyclizes relatively slowly. Cyclopropylcarbinyl halides and optically active alkyl halides have also seen service as probes (the former

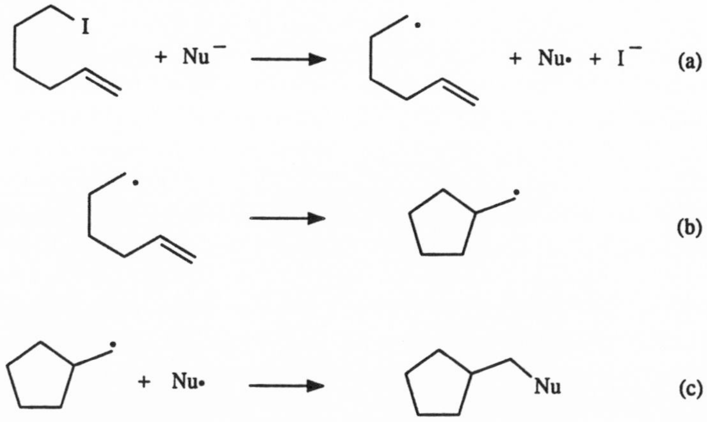

Figure 1. SET formulation of a nucleophilic substitution reaction with a mechanistic probe halide.

gives a radical that ring opens and the latter gives a radical that racemizes), but both provide less conclusive tests for radical intermediates than the hexenyl halides. Cyclopropylcarbinyl halides apparently can give ring-opened substitution products even when no "free" intermediate species is formed.[1] Because halide ion is liberated in nucleophilic substitutions, the optically active alkyl halides can racemize by simple halide attack.

Many of the reported alkyl halide mechanistic probe studies have been qualitative in nature. The products from reactions of the probes with nucleophiles were identified, and, when rearranged products were detected, mechanisms involving ET commonly were formulated. The logic for such a conclusion is that the detection of rearranged products from the probe implicates a radical intermediate and a radical intermediate is produced from one-electron reduction of an alkyl halide, which is a dissociative event. However, the implication of a radical intermediate with a lifetime adequate to permit the rearrangement is not sufficient proof that an ET reaction between the nucleophile and the alkyl halide occurred; for the mechanistic conclusion to be valid, one must establish that other pathways for formation of radicals from the alkyl halide do not exist. In fact, other pathways to radicals are possible, and, as we shall see, these reactions are the predominant pathways for radical formation in most mechanistic probe studies.

Alkyl halide mechanistic probe studies of nucleophilic substitution reactions became increasingly popular in the early 1980s. Both positive and negative results concerning the intermediacy of radicals were reported, and, when one allows for the fact that the negative results tended to be terminal studies whereas the positive results tended to be repeated and reported more than once, the frequencies of the two were comparable. Nevertheless, reactions of the probes with a number of nucleophiles and bases did give rearranged products, and often SET mechanisms were postulated to account for their formation. The novel concept that classical substitution and addition reactions actually occurred by electron transfer pathways was catchy, and SET pathways were the subject of reports in a chemical equivalent of the lay press, *Chemical & Engineering News*.[2] The "anionic" reagents tested with mechanistic probes included enolates,[3] alkoxides,[4] cuprates,[5] thiolates,[6] alkyllithiums,[7] stannyl anionoids,[1,8] amide bases,[9] and metal hydrides,[10] including reducing agents such as $LiAlH_4$. Some advocates of SET reactions concluded that SET reactions were a major component of organic reactions that had previously been overlooked.

It is instructive to consider just what is involved in a typical SET reaction pathway. Figure 1 presents a formulation of an S_N2 reaction involving a typical alkyl halide mechanistic probe, 6-iodo-1-hexene, that occurs by an SET pathway. The initial step is ET from the nucleophile to the alkyl halide; dissociative reduction of the alkyl halide gives a radical and a halide anion. The probe radical then rearranges, and the rearranged radical couples with Nu· (the radical formed by oxidation of the nucleophile) to give the observed rearranged products. The rearrangement, with a rate constant of about 10^5–10^6 M^{-1} s^{-1}, is slow in comparison to diffusional events, so the probe radical cannot react in a solvent cage but must instead first diffuse away from Nu· to give a free species. Thus, the SET pathways for reaction of a nucleophile with an alkyl halide involving long-lived radicals are not similar to cage reaction schemes, nor are they inherently related to the "electron shift" and similar formalisms[11] for interpreting a concerted nucleophilic addition or substitution reaction in terms of the potentials of the reactants. Some confusion exists concerning the latter point owing to the similarity of the terms.

The ET event in the SET pathway in Figure 1 was seldom discussed in detail, but it is clear that it was most commonly regarded as an "outer- sphere" ET process. That is, the reduction was envisioned to occur by an electron hop from the nucleophile to the alkyl halide rather than by an "inner-sphere" process, which would involve a bond formation followed by a homolysis step. Outer-sphere ET reactions can be (but seldom were) treated quantitatively by Marcus theory[12]; specifically, the difference in potentials between the oxidant and reductant and the reorganizational energy required for the ET process can be used to calculate the rate constant for ET. We shall return to this point later, but for now one may note that Marcus theory calculations apply to reversible ET reactions, and the dissociative nature of the reduction of an alkyl halide can confuse the issue. Furthermore, oxidation potentials for many of the nucleophiles studied were not available.

3. THE UNAPPRECIATED REACTION IN MECHANISTIC PROBE STUDIES

Our group had applied alkyl halide mechanistic probes in studies of reactions of stannyl anionoids conducted in the late 1970s and early 1980s. Based on this experience, we were not confident that mechanistic conclusions based primarily on the detection of rearranged products were entirely sound. The major shortcoming in many proposed reaction schemes involved the absence of kinetic information. The rate constants for rearrangements of the probe radicals either were known or could be measured by relatively simple methods that were well established, and other radical reactions that were postulated in the often complex SET schemes, such as reactions of radicals with solvent, also could be subject to kinetic studies. Thus, it appeared that the probe studies could be evaluated in a quantitative manner that might result in conclusions concerning the extent of ET. Instead, quite complex mechanisms often were postulated with little kinetic support, and, at times, kinetically untenable steps were included in the mechanisms.

For example, a radical–radical coupling reaction (step c in Figure 1) often was presented as the final step in an SET mechanism for a substitution reaction. However, for the simple alkyl radicals involved in mechanistic probe studies, radical–radical reactions typically occur at the diffusion limit and are notoriously unselective.[13] If ET processes were the only source of R· and Nu·, then equal amounts of R· and Nu· would be produced, and reaction between two R· radicals to give coupling and disproportionation products would be expected to occur with a velocity equal to that for formation of R-Nu, except in the special case in which Nu· is a stable radical that can accumulate to significant concentrations. However, researchers were not reporting detection of the expected coupling and disproportionation products. Another common proposal in SET mechanisms of metal hydride reactions was that a probe radical could react with solvent, typically an ethereal solvent such as tetrahydrofuran (THF), faster than it rearranged; this seemed to be highly unlikely in view of the fact that the radical rearrangements were reasonably fast.

More subtle kinetic evaluations also were inconsistent with some SET schemes. For example, in some schemes coupling between R· and Nu· was postulated as the source of unrearranged product R-Nu. With this assumption, one could claim that SET was the major reaction pathway even when limited amounts of rearranged products were formed. However, if radical–radical coupling is to be the source of unrearranged product, then it must compete with the probe rearrangement. In order to trap 50% of the 5-hexenyl radical before rearrangement, the pseudo-first-order rate constant for coupling at 25°C must be 2×10^{-5} s^{-1}. Because the second-order rate constant for diffusion-controlled radical–radical coupling in typical organic solvents at 25°C is about 5 $\times 10^9$ M^{-1} s^{-1},[13] this would require a concentration of Nu· of 4×10^{-5} M. In the

SET scheme, R· and Nu· are produced in equal amounts, and if their concentrations reached 4×10^{-5} M, the velocity of the coupling reaction would be about $8 \ M \ s^{-1}$. This means that a reaction of 0.1 M substrates would be complete in less than 0.1 s, but the reactions typically occurred over a period of minutes or hours.

Our apprehension about the details of the SET schemes solidified in the summer of 1984, and our group initiated a program aimed at placing mechanistic probe studies on a more quantitative basis. The original intent was to collect as much kinetic information as possible in order to evaluate the significance of the various radical reactions that might be involved, and the hope was that we could eventually state the extent of the ET processes. The kinetics are discussed in the next section, but suffice it to state here that, although the data accumulated, the picture did not clarify. Instead, it became increasingly apparent that some point was being overlooked in the interpretation of mechanistic probe studies.

The missing factor was a failure to appreciate the importance of a facile pathway for isomerization of the alkyl halide probes to rearranged alkyl halides. This isomerization occurs by the sequence shown in Figure 2. An initiation reaction of some type gives the probe radical that rearranges. The rearranged radical can then abstract a halogen atom from a probe molecule to give the rearranged alkyl halide and another probe radical. The rearrangement and atom transfer steps are the propagation steps of a radical chain reaction. If both steps are fast, a small amount of radical initiation will result in a large amount of

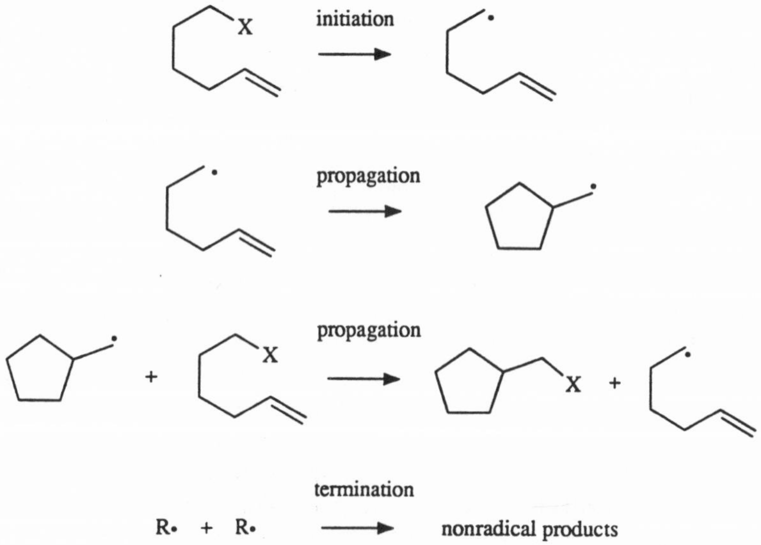

Figure 2.   The radical chain isomerization sequence.

isomerized alkyl halide. In a mechanistic probe study, the intrusion of the radical chain isomerization pathway can result in isomerization of the probe in competition with the nucleophilic substitution reaction. When this happens, the rearranged alkyl halide can simply react with the nucleophile by a conventional polar process to give rearranged substitution products. Thus, in principle, large amounts of rearranged products could be detected in a probe study as long as some radical initiation process occurred; ET between the nucleophile and alkyl halide is not necessarily required to account for rearranged products.

It should be emphasized that the radical chain sequence in Figure 2 was not a new conception. In the mid-1960s, Brace reported the sequence as a synthetic method for isomerizing acyclic alkyl iodides to cyclic compounds.[14] The halogen atom transfer step in Figure 2 is one step in the well-known inter-molecular addition of alkyl halides, especially polyhalogenated methanes, to olefins, and it was documented by Kharasch in the 1940s in the addition reactions of α-bromo esters to olefins.[15] The results from electron spin resonance (ESR) and chemically induced dynamic nuclear polarization (CIDNP) studies conducted since the 1960s of reactions of alkyllithium reagents with alkyl bromides and iodides required that halogen atom exchange be fast,[16] and experiments directed at measuring the ratios of rate constants for radical recombinations and radical heats of formation were designed around the fact that rapid iodine atom transfer would permit an equilibration of radicals.[17] In fact, isomerized alkyl halides were observed in some mechanistic probe studies conducted by Ashby's group, and their production was ascribed to the atom transfer step in Figure 2.[10e-g] The point is that, although the radical chain isomerization sequence was known, its significance was overlooked.

Our interest in the radical chain isomerization process arose late in 1985, and preliminary kinetic studies that confirmed the importance of the sequence in probe studies were completed within a few weeks. Although not immediately appreciated, a key event in directing our attention to the atom transfer step was a discussion between Professor Dennis Curran and the author that took place earlier that year and that focused on Curran's imaginative applications of the radical cyclization–atom transfer reactions in synthesis.[18] Curran and co-workers also realized the mechanistic significance of their work, and publications that contained the caveat for mechanistic probe studies were independently submitted by the two groups.[19,20] Eventually, the common interests in the rates of atom transfer reactions led to collaborative efforts of the two groups, including a critical evaluation of mechanistic probe studies.[21]

4. KINETICS OF REACTIONS OCCURRING IN MECHANISTIC PROBE STUDIES

The formation of rearranged products in alkyl halide mechanistic probe studies clearly indicates that radicals are formed in a number of reactions of alkyl

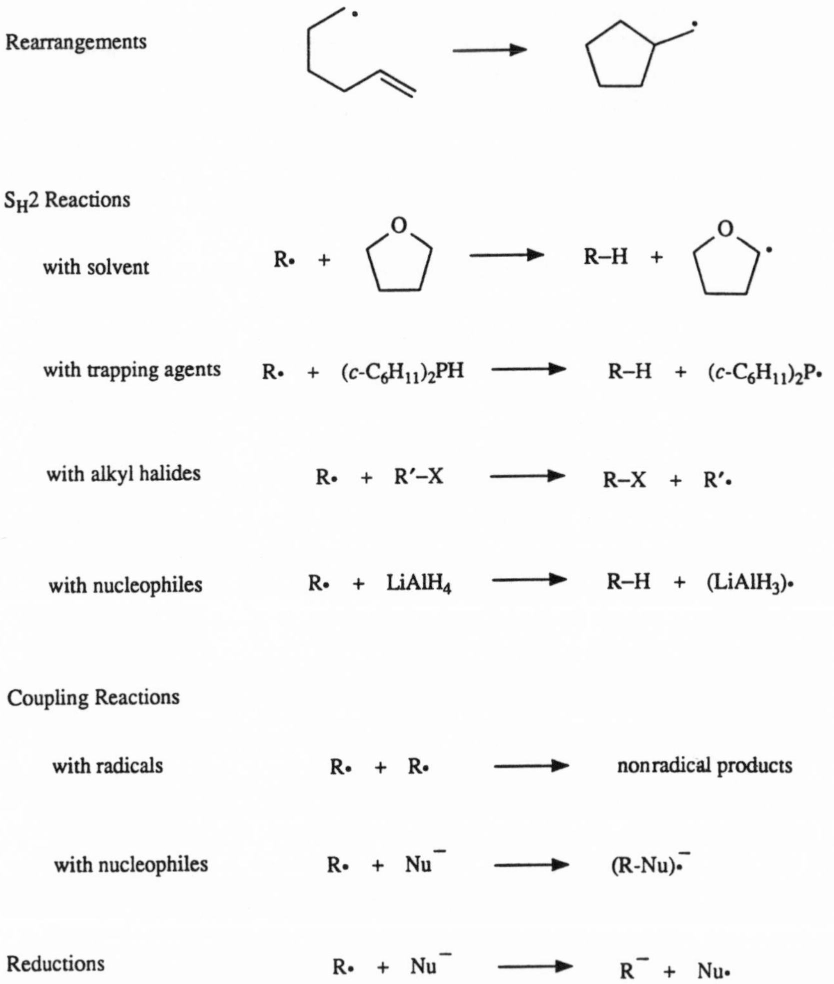

Rearrangements

S_H2 Reactions

with solvent   R• +   ⟶   R–H +

with trapping agents   R• + $(c\text{-}C_6H_{11})_2PH$ ⟶   R–H + $(c\text{-}C_6H_{11})_2P•$

with alkyl halides   R• + R'–X   ⟶   R–X + R'•

with nucleophiles   R• + $LiAlH_4$ ⟶   R–H + $(LiAlH_3)•$

Coupling Reactions

with radicals   R• + R•   ⟶   nonradical products

with nucleophiles   R• + Nu^- ⟶ $(R\text{-}Nu)•^-$

Reductions   R• + Nu^- ⟶ R^- + Nu•

Figure 3.   Possible radical reactions in mechanistic probe studies.

halides with nucleophiles and bases, but the implication of a radical intermediate does not reveal whether radicals are formed in the reaction of interest
or in a competing reaction sequence. In order to answer the latter question, one
must evaluate the kinetics of the possible reactions. There is now an adequate
collection of rate constants for reactions of radicals at or near room temperature
to permit one to understand what the major reaction pathways are in most of
the reported mechanistic probe studies. For some reductions of alkyl halides by
metal hydride reducing agents, sufficient kinetic information exists to allow

one to calculate a maximum limit for the amount of ET in a mechanistic probe study that gave rearranged products.

In considering radical kinetics, it is convenient to separate the reactions that are or might be important in mechanistic probe studies into several groups. These are collected in Figure 3 along with representative examples. In addition to several known radical reactions, this collection contains some radical reactions that have been proposed in SET schemes but have not been observed. Unknown reactions create a problem in kinetic analyses, but in some cases one can set a limit on the rate constant of the unknown reaction by observing that the reaction did not occur when other reactions with known rate constants did occur.

The rate constants for rearrangements of the probe radicals are the best understood.[22] In most cases, the rearrangement involves the isomerization of one radical with no special stabilizing groups into another radical with no stabilizing groups. Little charge development is expected in the transition state for such a rearrangement, and, therefore, there should be little if any solvent effect on the rate of the reaction. Indeed, the popular "radical clock" kinetic method,[22] in which the rate constant of a reaction of interest is "timed" against a radical rearrangement, often is applied with the implicit assumption that the rearrangement rate constant can be transferred from one solvent to another.

Among the various radical rearrangements that have been applied in mechanistic probe studies, the 5-hexenyl radical cyclization is one of the slowest. This cyclization has a rate constant of 2.2×10^5 s^{-1} at 25°C.[23] Alkyl-substituted analogs of 5-hexenyl have similar rate constants.[22a] Faster radical rearrangements are known, and many of these are well calibrated. However, because most of the other possible radical reactions in Figure 3 occur with pseudo-first-order rate constants at 25°C that are less than 1×10^5 s^{-1}, the 5-hexenyl radical cyclization provides a convenient limiting case for kinetic analyses.

The S_H2 reactions (bimolecular homolytic substitution) in Figure 3 are the most important for understanding mechanistic probe studies. These include hydrogen atom abstractions from solvents, which appear in several SET schemes as pathways for formation of reduced products (R-H) and for formation of protium-substituted products when deuterated metal hydride reducing agents such as LiAlD$_4$ are employed. Similarly, trapping agents that have been employed in some probe studies, such as dicyclohexylphosphine (DCPH) and 1,4-cyclohexadiene (CHD), can react with radicals by hydrogen atom transfer. The halogen atom transfer reactions of a radical with an alkyl halide are the crucial steps in the radical chain isomerization pathways that can result in isomerization of a mechanistic probe into a rearranged alkyl halide. S_H2 reactions of radicals with nucleophiles, such as hydrogen abstraction from LiAlH$_4$, have figured prominently in proposed SET pathways.

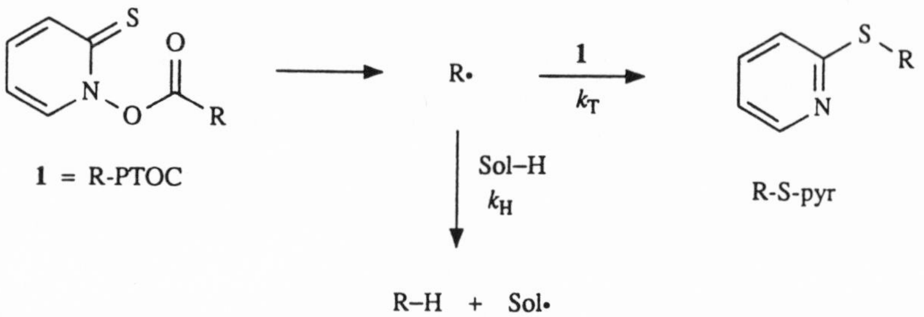

Figure 4. Competition experiment for measuring k_H.

Most of the mechanistic probe studies of interest here have been conducted in the ethereal solvents THF and diethyl ether (Et$_2$O). Owing to the large and unchanging concentration of the solvent, rate constants for reactions of radicals with solvent are pseudo-first-order. Kinetic studies of reactions of simple alkyl radicals with THF were determined by a competition method in which "self-trapping" of the radicals by N-hydroxypyridine-2-thione esters (PTOC esters, **1**)[24] and hydrogen atom transfer from solvent were the two available reaction channels (Figure 4). In the self-trapping reaction, alkyl radical adds to the thione sulfur of a precursor molecule to give, ultimately, an alkyl pyridyl sulfide (R-S-pyr), whereas the hydrogen transfer reaction gives a hydrocarbon product (R-H). The rate constants for hydrogen atom abstraction (k_H) were calculated from the known[25] rate constant for self-trapping (k_T), the concentration of PTOC ester and the product ratio:

$$k_H = k_T \, ([\text{R-H}]/[\text{R-S-pyr}]) \, [\text{R-PTOC}]_m \qquad (1)$$

The results for alkyl radicals[26] are given in Table 1.

Rate constants for reactions of the trapping agents DCPH and CHD were determined by the radical clock method.[22] PTOC esters **2** and **3** were used as the precursors of the 5-hexenyl radical and the 2,2-dimethyl-3-butenyl radical,

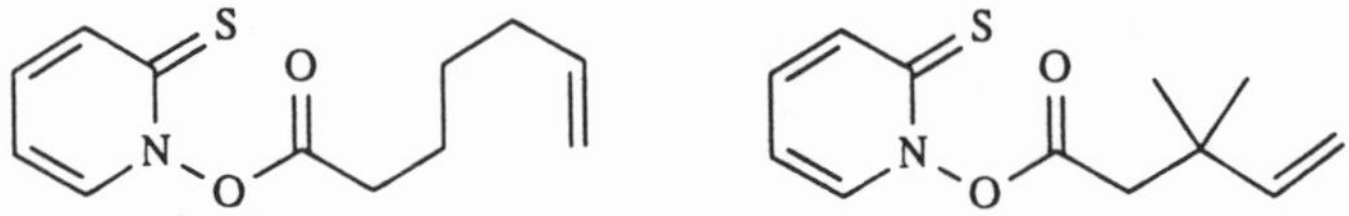

respectively. These radicals were either trapped by the hydrogen atom donor or rearranged, and the rearranged radical then reacted with the hydrogen donor. Figure 5 shows the scheme for reactions with PTOC ester **3**. From the yields

Table 1. Rate Constants for Hydrogen Atom Transfer Reactions

Donor	Radical	Temp. ($°C$)	k	Ref.
Ether[a]	Octyl	22	$1.1 \times 10^3 \, s^{-1}$	26b
THF[b]	Octyl	22	$6 \times 10^3 \, s^{-1}$	26b
	Cyclopentylmethyl	50	$6 \times 10^3 \, s^{-1}$	26a
	1,1-Dimethyl-3-butenyl	50	$2 \times 10^3 \, s^{-1}$	26a
CHD[c]	5-Hexenyl	50	$2.3 \times 10^5 \, M^{-1} \, s^{-1}$	26a
	2,2-Dimethyl-3-butenyl	50	$4.8 \times 10^5 \, M^{-1} \, s^{-1}$	26a
	Ethyl	27	$5.8 \times 10^4 \, M^{-1} \, s^{-1}$	27
	tert-Butyl	27	$9.4 \times 10^3 \, M^{-1} \, s^{-1}$	27
DCPH[d]	5-Hexenyl	50	$7 \times 10^5 \, M^{-1} \, s^{-1}$	26a
	2,2-Dimethyl-3-butenyl	27	$1 \times 10^6 \, M^{-1} \, s^{-1}$	26a

[a]Diethyl ether.   [b]Tetrahydrofuran.   [c]1,4-Cyclohexadiene.   [d]Dicyclohexylphosphine.

of the two hydrocarbon products, the concentrations of the trapping agents and the known rate constants for the rearrangements, the rate constants for trapping (k_H) were calculated by the following equation:

$$k_H = k_r \, ([R\text{-}H]/[R'\text{-}H]) \, [Y\text{-}H]_m^{-1} \qquad (2)$$

where k_r is the rate constant for rearrangement, [R-H] is the yield of unrearranged hydrocarbon, [R'-H] is the yield of rearranged hydrocarbon, and [Y-H]$_m$ is the mean concentration of the trapping agent during the reaction. The rate constants for these reactions[26a] are also given in Table 1. The rate constants for reaction of CHD with the simple ethyl and *tert*-butyl radicals[27] are in reasonable agreement with the values determined by the radical clock method.

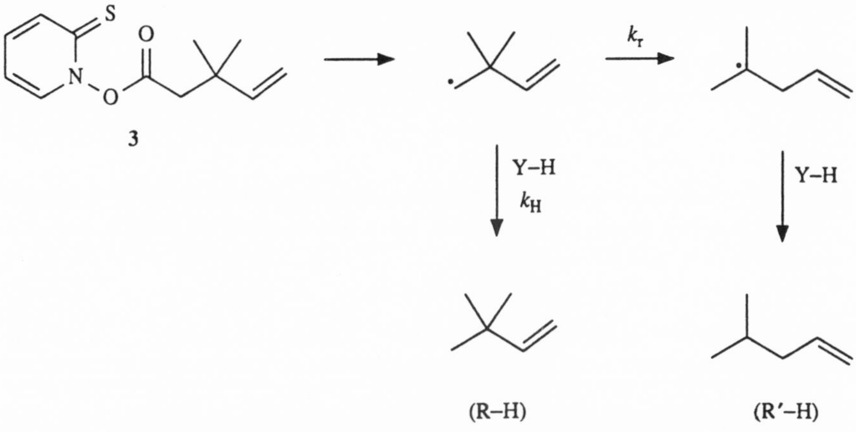

Figure 5. Radical clock method for measuring k_H.

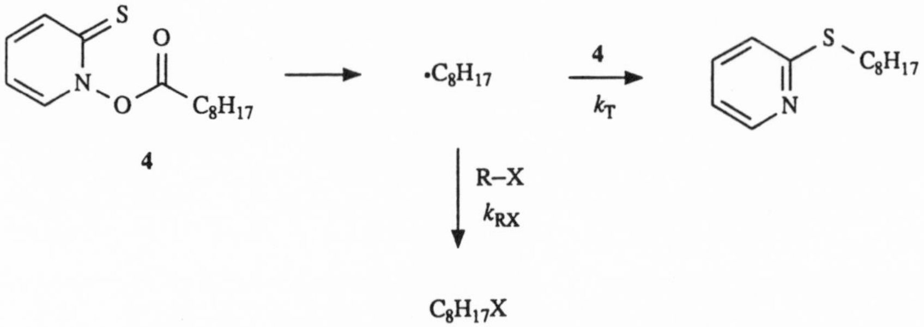

Figure 6. PTOC trapping method for measuring k_{RX}.

Rate constants for halogen atom abstraction from a variety of alkyl halides by octyl radical, selected as a representative primary radical, were determined. The PTOC ester **4** served as the source of the alkyl radical and as the radical trapping agent with a known[25] reaction rate constant. Reaction of **4** in the presence of an alkyl halide gave octyl pyridyl sulfide from self-trapping and octyl halide from reaction with the alkyl halide RX (Figure 6). Rate constants for halogen atom transfer (k_{RX}) were calculated from the following competition expression:

$$k_{RX} = k_T \, ([\text{Oct-X}]/[\text{Oct-S-pyr}]) \, ([4]/[\text{R-X}])_{eff} \qquad (3)$$

where k_T is the rate constant for self-trapping, ($[\text{Oct-X}]/[\text{Oct-S-pyr}]$) is the observed ratio of the products, and ($[4]/[\text{R-X}])_{eff}$ is the effective concentration ratio of the reagents during the reaction. The initial studies[20] were conducted at 50°C; later studies[26b] of the reactions of octyl radical with iodoethane, 1-iodobutane, and iodocyclohexane, conducted at 22°C, were more precise. The kinetic data are collected in Table 2. It is noteworthy that the self-trapping reaction of the octyl radical by its PTOC ester was used as the competition reaction in both the halogen atom transfer reactions and the reactions with ethereal solvents. Therefore, the ratios of rate constants for the two measured reactions are relatively accurate because any error in the rate constant for the self-trapping reaction cancels by division.

For the alkyl iodides, rate constants for iodine abstraction could be determined by another competition method.[20] A mixture containing the iodine donor (R-I) and octyl bromide in benzene was treated with an insufficient amount of Bu₃SnH. Tin hydride reduces alkyl halides via a radical chain reaction, the propagation reactions of which are shown in Figure 7. The Bu₃Sn· radical reacted predominantly with R-I, but some octyl radical was formed by reaction of Bu₃Sn· with Oct-Br. The octyl radical thus formed was either reduced to Oct-H by reaction with the tin hydride or converted to Oct-I by reaction with the iodine donor R-I. Of course, the Oct-I formed in the latter case is subject to

Table 2. Rate Constants for Halogen Atom Transfer to Octyl Radical[a]

Halogen Donor	Method[b]	Temp. (°C)	$k \ (M^{-1} \ s^{-1})$
$(CH_3)_3CI$	Tin hydride	50	3×10^6
$(CH_3)_2CHI$	Tin hydride	50	6×10^5
	PTOC	50	9×10^5
$c\text{-}C_6H_{11}I$	Tin hydride	50	5×10^5
	PTOC	50	5×10^5
	PTOC	22	2.6×10^5
CH_3CH_2I	Tin hydride	50	2×10^5
	PTOC	50	3×10^5
	PTOC	22	1.4×10^5
$CH_3(CH_2)_3I$	PTOC	22	1.1×10^5
$(CH_3)_3CBr$	PTOC	50	5×10^3
$(CH_3)_2CHBr$	PTOC	50	1.2×10^3
$c\text{-}C_6H_{11}Br$	PTOC	50	0.8×10^3
$CH_3(CH_2)_3Br$	PTOC	50	0.6×10^3
$(CH_3)_3CCl$	PTOC	50	0.6×10^3

[a]From refs. 20 and 26b.
[b]The competition reactions were trapping by Bu_3SnH (tin hydride) or trapping by the PTOC ester precursor (PTOC).

tin hydride reduction, but, by using a large excess of R-I and Oct-Br, we could minimize this reaction by a concentration effect. The rate constants for reactions of simple radicals with Bu_3SnH (k_H) are well established,[28] so the rate constants for iodine atom transfer could be calculated from the following equation:

$$k_{RI} = k_H \ ([\text{Oct-I}]/[\text{Oct-H}]) \ ([Bu_3SnH]/[\text{R-I}])_{\text{eff}} \qquad (4)$$

where $([\text{Oct-I}]/[\text{Oct-H}])$ is the observed ratio of octyl iodide to octane and $([Bu_3SnH]/[\text{R-I}])_{\text{eff}}$ is the effective ratio of the reagents. Given the level of approximation in the study, reasonable agreement was obtained for the rate constants measured by the two methods. Table 2 contains the results.

$$R\bullet + Bu_3SnH \longrightarrow R\text{–}H + Bu_3Sn\bullet$$

$$R\text{–}I + Bu_3Sn\bullet \longrightarrow R\bullet + Bu_3SnI$$

$$Oct\text{–}Br + Bu_3Sn\bullet \longrightarrow Oct\bullet + BuSnBr$$

$$Oct\bullet + Bu_3SnH \longrightarrow Oct\text{–}H + Bu_3Sn\bullet$$

$$Oct\bullet + R\text{–}I \longrightarrow Oct\text{–}I + R\bullet$$

Figure 7. Propagation reactions in the tin hydride method for measuring k_{RX}.

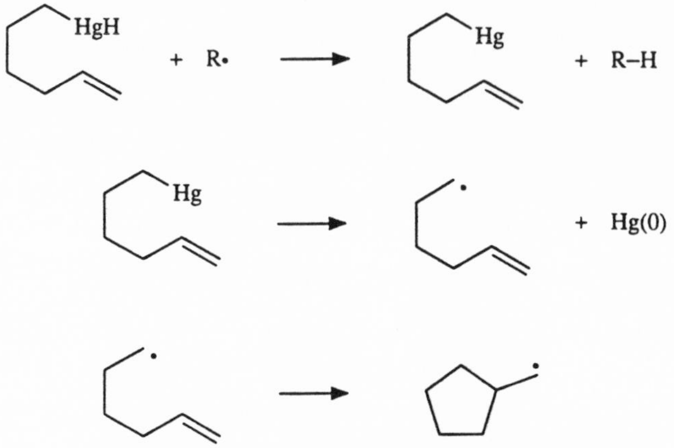

Figure 8.   Propagation reactions in the Russell and Guo study.

The S_H2 reaction of an alkyl radical with a metal hydride to give an alkane and some type of radical from the metal hydride has appeared in a number of SET schemes that purported to detail the mechanisms of reactions of alkyl halides with metal hydrides like LiAlH$_4$. In fact, according to published schemes, some would believe that this reaction is so fast for LiAlH$_4$ that it can compete with rearrangements of probe radicals like 5-hexenyl. Our group has not attempted to measure rate constants for these reactions because data in the literature showed that the reactions were slow for the archetypal metal hydrides, NaBH$_4$ and LiAlH$_4$.

Russell and Guo[29] investigated the reaction sequence shown in Figure 8. 5-Hexenylmercury(II) chloride at constant concentration was allowed to react with NaBH$_4$ at various concentrations, and the ratios of the products (1-hexene and methylcyclopentane) were determined. Alkylmercury(II) halides are reduced by NaBH$_4$ to alkylmercury(II) hydrides (RHgH). The mercury hydride is a reactive hydrogen atom donor for an alkyl radical; the reaction gives an alkane and an alkylmercury(I) radical that decomposes by loss of Hg(0) to give an alkyl radical. In their study, Russell and Guo found that the ratio of 1-hexene to methylcyclopentane was constant when the concentration of NaBH$_4$ was varied. In other words, the constant concentration RHgH from the RHgCl trapped some 5-hexenyl radical in competition with the cyclization reaction, but no additional trapping was apparent that could be ascribed to the NaBH$_4$. A conservative interpretation of their results requires that the rate constant for reaction of 5-hexenyl radical must be less than 1×10^4 M^{-1} s^{-1} at 30°C; the rate constant probably is less than 1×10^3 M^{-1} s^{-1}.[29]

$$(CH_3)_3COOC(CH_3)_3 \xrightarrow{h\nu} 2 \ (CH_3)_3CO\cdot$$

$$(CH_3)_3CO\cdot \ + \ LiAlH_4 \longrightarrow (CH_3)_3COH \ + \ (LiAlH_3)\cdot$$

$$(LiAlH_3)\cdot \ + \ RX \longrightarrow LiAlH_3X \ + \ R\cdot$$

$$R\cdot \ + \ LiAlH_4 \longrightarrow R–H \ + \ (LiAlH_3)\cdot$$

$$R\cdot \ + \ Sol–H \longrightarrow R–H \ + \ Sol\cdot$$

$$R\cdot \ + \ R\cdot \longrightarrow \text{coupling and disproportionation}$$

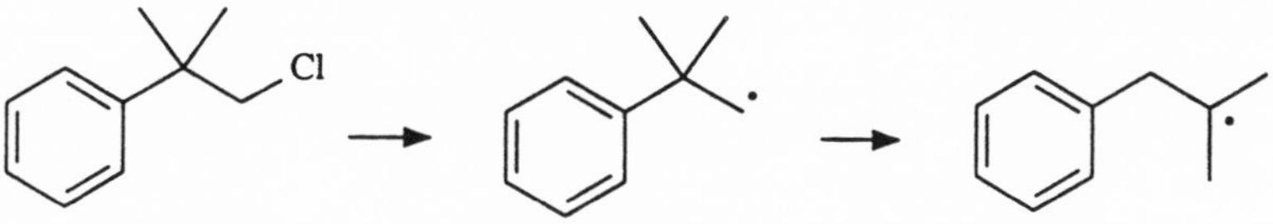

5   6   7

Figure 9. Details of the Beckwith and Goh study.

The rate constant for reaction of $LiAlH_4$ with a representative alkyl radical can be derived from data available in the literature. Beckwith and Goh[30] studied the photochemical reactions of alkyl halides with $LiAlH_4$ in ether solutions containing di-*tert*-butyl peroxide, which resulted ultimately in reduction of the alkyl halides to alkanes (Figure 9). In these reactions, photolysis of the peroxide gave the *tert*-BuO· radical, which, from ESR studies,[31] is known to react with $LiAlH_4$ to give a species that can be formulated as $(LiAlH_3)\cdot$. This species is known to react rapidly with alkyl halides (including alkyl chlorides) at 205 K to give alkyl radicals.[31] Qualitatively, the sequence of reactions described to this point provides some support for the notion presented in some SET schemes that an alkyl radical can abstract hydrogen from $LiAlH_4$. However, one of the alkyl chlorides studied by Beckwith and Goh was neophyl chloride (**5**). Reduction of this chloride in a solution that initially contained 1 M $LiAlH_4$ gave *tert*-butylbenzene from reduction of radical **6** and (2-methylpropyl)benzene from reduction of radical **7** in a 4 : 1 ratio. The rearrangement of **6** to **7**, the neophyl rearrangement, occurs slowly ($k_r = 750 \ s^{-1}$ at 25°C).[32] If one assumes that the only source of hydrogen for reduction in these experiments was $LiAlH_4$, then the rate constant for the hydrogen atom abstraction (k_{LAH}) would be given by the relation

$$k_{LAH} = (750 \ s^{-1}) \ (4/1) \ (1 \ M)^{-1} = 3000 \ M^{-1} \ s^{-1} \qquad (5)$$

where $750 \ s^{-1}$ is the rate constant for the rearrangement, 4 : 1 is the ratio of unrearranged to rearranged products, and 1 M is the concentration of $LiAlH_4$.

$$R–X \longrightarrow R\cdot$$

$$R\cdot + Nu^- \longrightarrow (R–Nu)\cdot^-$$

$$(R–Nu)\cdot^- + R–X \longrightarrow R–Nu + R\cdot + X^-$$

Figure 10. The $S_{RN}2$ reaction mechanism.

The derived rate constant for hydrogen abstraction from LiAlH$_4$ not only is small but also is almost certainly only an *upper* limit, because the reaction of neophyl radical with ether ($k = 1000$ s^{-1}) must have accounted for formation of some of the unrearranged product and because the high UV flux in the experiment probably resulted in high concentrations of radicals that could disproportionate.

Radical coupling with a nucleophile is another type of reaction that has been postulated in SET schemes formulated to account for the results of probe studies. The coupling reaction would produce a radical anion that could then reduce an alkyl halide to give a neutral product, a radical, and a halide ion. These reactions comprise a radical chain reaction sequence that is known to be important in the $S_{RN}2$ reactions studied by Russell, Kornblum, and Bunnett (Figure 10).[33] Proponents of SET mechanisms have on occasion incorporated the radical–nucleophile coupling step and subsequent ET step into the proposed pathways for product formation, apparently without realizing that if these reactions occurred then the formation of rearranged products would be ex-plained by the $S_{RN}2$ pathway rather than the purported SET pathway.

In general, mechanistic probe studies have employed localized anionic species as the nucleophiles and probes that give rise to localized radicals. This is an important difference between the probe studies and actual $S_{RN}2$ reactions in which the anionic species or the radical (or both) are delocalized. When some form of delocalization is possible for the radical anion intermediate formed by addition of the radical to the nucleophile, the reaction is relatively facile. However, when there is no delocalization in the purported radical anion, the adduct would be an especially high-energy intermediate containing an electron in a σ^* orbital, and it would form only slowly if at all.

Kinetic studies conducted by Russell demonstrate the importance of delocalization for formation of a radical anion by coupling of a radical with a nucleophile. The delocalized 2-nitro-2-propyl radical adds to delocalized anions and localized anions with about the same rate constants.[34a] However, localized alkyl radicals, which readily add to delocalized nitronate anions, do not react measurably with localized anions.[34b] Approximate rate constants for addition of localized radicals to various delocalized nucleophiles at 35°C are in the range of $4–20 \times 10^3$ M^{-1} s^{-1}.[34c] With the assumption that localized alkyl radicals will react with localized anions several orders of magnitude less rapidly than they do with delocalized anions, one predicts a rate constant for the addition step that is much too small to be important for mechanistic probe studies. Even the relatively slow reaction of a simple radical with an ethereal solvent would be expected to overwhelm the reaction of the radical with the localized nucleophile.

Radical–radical reactions, including couplings and atom transfer processes, have been invoked in a number of SET reaction pathways as the final, product-forming step. This is probably the most serious shortcoming of SET schemes in general. The radical–radical reactions often appear to have been included simply because they provided a convenient route to complete a scheme once large numbers of radicals had been predicted by the earlier steps of the scheme, and discussions of SET pathways have commonly failed to consider the kinetics of radical–radical reactions, the expected products from radical–radical reactions or the concentrations of radical intermediates.

Unlike the case for the radical–molecule reactions considered to this point, the relative velocities of radical–radical reactions cannot be established for a general case but must instead be determined for each specific reaction. This, of course, results because the processes are second-order in radical concentrations, and one cannot calculate a pseudo-first-order rate constant without knowing the radical concentrations. In mechanistic probe studies, it is often possible to estimate the average radical concentration in the reaction from the overall velocity of the conversions that might involve radicals and the rate constants of the fast radical reactions that can occur. An example of the method is now given.

Let us assume that a mechanistic probe study at room temperature employing initially 0.1 M alkyl iodide probe proceeds with a half-life of 1 h and results in formation of rearranged products in 10% yield. In the first half-life, the velocity of the reactions that gave rearranged products was 0.005 M h^{-1} or about 1×10^{-6} M s^{-1}. Based on the relative rate constants we have considered, radical rearrangement will be the fastest radical process and iodine atom transfer will be the next fastest radical process; together these two steps comprise the radical chain isomerization reaction that will result in isomerized products. All steps in a chain reaction have the same velocity, so the velocity of the chain process (v_{isom}) is given by

$$v_{isom} = k_{RX} [R\cdot] [R\text{-}X]_m \qquad (6)$$

where k_{RX} is about 1×10^5 M^{-1} s^{-1}, and $[R\text{-}X]_m$, the average concentration of the halide, is 0.075 M in this example. From equation 6, the radical concentration ($[R\cdot]$) is about 1×10^{-10} M. For simple alkyl radicals in common organic solvents, radical–radical reactions occur under diffusion control with a rate constant of about 5×10^9 M^{-1} s^{-1} at 25°C.[13] Thus, one can now calculate the velocity of radical–radical reactions for our example as shown here:

$$v_{(R \cdot R \cdot)} = (5 \times 10^9 \text{ M}^{-1} \text{ s}^{-1}) [R \cdot]^2 \tag{7a}$$

$$= (5 \times 10^9 \text{ M}^{-1} \text{ s}^{-1})(1 \times 10^{-10} \text{ M})^2 = 5 \times 10^{-11} \text{ M s}^{-1} \tag{7b}$$

$$k'_{R \cdot} = v_{(R \cdot R \cdot)}/[R \cdot] = 0.5 \text{ s}^{-1} \tag{8}$$

Comparing the velocity of the radical–radical reactions in equation (7b) to the velocity of the chain reactions (1×10^{-6} M s^{-1} in this example) shows that radical–radical couplings could not be important. Alternatively, the velocity in equation (7) can be divided by [R·] to give a pseudo-first-order rate constant for radical–radical reactions [$k'_{R \cdot}$ in equation (8)] that can be compared to the first-order and pseudo-first-order rate constants in Tables 1 and 2.

As noted previously, the average concentration of radicals will differ from one study to another. It is of interest, therefore, to turn the question of radical concentrations around and ask "What concentration of radicals would be required such that radical–radical reactions predominate in a mechanistic probe study?" The answer to this question will permit a calculation of the overall velocity of a mechanistic probe study in which radical–radical coupling could be the major source of products.

For convenience, consider the case in which 50% of the radicals couple and 50% react with a primary alkyl iodide at 0.1 M concentration. The equality in the two reaction velocities gives the following equation:

$$5 \times 10^9 \text{ M}^{-1} \text{ s}^{-1} [R \cdot]^2 = 1 \times 10^5 \text{ M}^{-1} \text{ s}^{-1} (0.1)[R \cdot] \tag{9}$$

which leads to a radical concentration of 2×10^{-6} M. This radical concentration is within the range of concentrations that can be detected by ESR spectroscopy, so ESR or CIDNP experiments should indicate radicals if the SET pathway were utilized. However, when a radical concentration of 2×10^{-6} M is used in equation (7a), one sees that $v_{(R \cdot R \cdot)}$ would be about 0.02 M s^{-1}, and the half-life of the reaction would be only a few seconds. This exercise permits the following generalization for an alkyl iodide probe study: If the reaction half-life is minutes or more, SET reactions cannot be the major pathways for formation of radicals from the precursor.

Another radical reaction that is important when strong reducing agents react with an alkyl halide is reduction of the radical to an anion. This possibility has usually been overlooked, but it is important. Simple primary alkyl radicals are actually stronger oxidizing agents than their alkyl bromide precursors and might be stronger oxidizing agents than alkyl iodides, and the estimated intrinsic energy barrier for ET to simple radicals[35] is at most only about half of that estimated for alkyl halides.[36] The combination of these two features means that a reducing agent will reduce a primary alkyl radical faster than it reduces an alkyl bromide. This property is well known in electrochemistry, where cyclic voltammetry (CV) studies of alkyl halides often result in a single, two-electron wave for reduction, showing that reduction of the alkyl halide occurred at a potential more negative than that required to reduce the radical. For example in a recent report, Savéant et al.[35] found that 1-bromobutane, 2-bromobutane and even 1-iodobutane exhibited single, two-electron waves in CV studies.

Rate constants for radical reductions are not generally available at this time. In principle, they can be calculated by Marcus theory[12] when the reduction potentials and the reorganizational energies for the radicals and the reducing agents are known. Ongoing work in a few laboratories[35,37] suggests that the reduction potentials for several radicals will soon be known; for the simple radicals *tert*-butyl, *sec*-butyl and *n*-butyl, reduction potentials of −1.55, −1.45 and circa −1.35 V versus SCE have been reported.[35] The important point for those investigating possible SET pathways is that if the reagent of interest is a strong enough reducing agent to reduce an alkyl halide, it will probably also reduce an alkyl radical.

5. INTERPRETATION OF MECHANISTIC PROBE STUDIES

The various rate constants for radical reactions can be used to interpret or to predict the outcome of a mechanistic probe study. Most probe studies have employed alkyl iodides, and, as we have seen, iodine atom transfer is the fastest second-order radical reaction channel available unless the overall conversion is exceedingly fast and high concentrations of radicals are produced. Reaction with solvent THF is the next fastest second-order reaction when alkyl iodides are employed. The competition between reaction of a radical with an alkyl iodide to give R-I and reaction with an ethereal solvent like THF to give R-H obviously affects the product distribution, but the solvent reaction does not prevent the radical chain isomerization. For example, radical chain isomerizations of 6-iodo-1-hexene to (iodomethyl)cyclopentane were shown to be efficient in THF.[20] Apparently, the radical formed by hydrogen abstraction from THF simply abstracts an iodine atom from the alkyl iodide in another chain propagation step. Thus, for alkyl iodide probes, once a radical chain reaction has been initiated, radical chain isomerization of the probe will ensue; the sequence amplifies the radical initiation events, which might involve ET. A quantitative evaluation of the results of mechanistic probe studies is necessary if one wishes to determine the extent of radical initiation and thus the *maximum* possible amount of ET.

The above considerations lead to immediate qualitative predictions that can be verified. Specifically, if radical chain isomerization is the major route for radical formation in an alkyl iodide probe study, then one might expect to see little or no evidence of radical intermediates in experiments designed to prevent the chain reaction. Two possible experimental designs come to mind. In one, the atom transfer step could be avoided by changing the identity of the halogen atom from iodine, which is abstracted readily, to chlorine. Chlorine atom transfer from primary and secondary alkyl chlorides to octyl radical was too

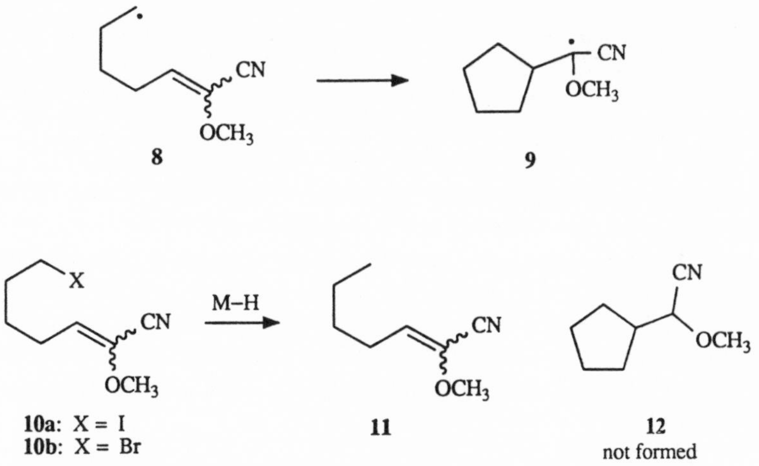

Figure 11. Details of the Park, Chung, and Newcomb study.

slow to measure by our methods, and the rate constants for these reactions must be less than that for reaction of *tert*-butyl chloride ($k = 600$ M^{-1} s^{-1} at 50°C).[20] In cases in which investigators have tested alkyl chloride probes in reactions with weakly reducing nucleophiles, rearranged products have not been found even though alkyl iodide probes did give rearranged products under similar reaction conditions. This is consistent with the prediction that radical chain isomerization of alkyl chlorides is not possible and the expectation that only a very few radical-producing reactions occur. One might point out that alkyl chlorides are more difficult to reduce than alkyl iodides and ET to alkyl chlorides is expected to be slower than ET to iodides, but the importance of this observation is not clear because alkyl chlorides will also react less rapidly in substitution reactions and therefore will be more persistent than alkyl iodides.

A more conclusive qualitative test was provided by an alkyl iodide probe that was not able to participate in a radical chain isomerization sequence (Figure 11). We have shown that cyclization of the 6-cyano-6-methoxy-5-hexenyl radical (8) to radical 9 is about 400 times faster than cyclization of its parent, 5-hexenyl radical.[38a] The cyclic radical 9 is a captodative stabilized radical,[38b] and thus it was expected to be too stable to abstract iodine from an alkyl iodide. When the alkyl halides 10 were employed as mechanistic probes in reactions with various metal hydrides, including NaBH$_4$ and LiAlH$_4$, acyclic product 11 was formed but no cyclic product 12 was detected.[39] The functionality in probes 10 is so far removed from the carbon–halogen bonds that one would not expect any difference between the reduction potential of iodide 10a and that of a simple primary iodide. Furthermore, radical 8 cyclizes hundreds of times faster than other probes that had given appreciable yields of cyclic products in reactions with metal hydrides, so any radical **8** that was formed from halides **10** would have cyclized. The reasonable conclusion is that the amount of radical formed from the halide probes was so small that the radical-derived products were not detectable; that is, the amount of radical initiation, possibly by SET, must have been less than 0.1%, a conservative estimate of our limit of detection by GC.[39]

Subsequent to our publication, Ashby's group has reported in a review and a communication studies of reactions of iodide **10a** with LiAlH₄.[40] These publications do not contain full experimental details, but it is apparent that in many reactions the Ashby group confirmed our observations that probe **10a** does not give any cyclic product in reactions with LiAlH₄. However, the Ashby group concluded that the test was "invalid" because they found that the *E/Z* isomer ratio in reduced product **11** was different from that in probe **10a** and because reduction of probe **10a** with LiAlD₄ gave traces of **11**-*d₀*. They also found cyclic products in LiAlH₄ reductions of 8-iodo-3-methyl-3-octene (**13**). Based on their observations, they postulated the sequence of events in Figure 12. The absence of cyclic products was attributed to the formation of intermediate **15** (which is both a radical and a radical anion at the same time) with the claim that the radical anion portion of the species somehow prevented the cyclization. The altered *E/Z* isomer ratio in the reduced product was rationalized by the formation of **16**, which was presumed to give a new *E/Z* ratio in an unexplained oxidation process. The results with iodide **13** were claimed to support the mechanistic speculations because **13** was considered to be equivalent to **10a** owing to the fact that the radical from **13** would give a tertiary radical upon cyclization.

One can dismiss the scheme in Figure 12 for a variety of reasons. The most important is that the steps have little precedent beyond speculations contained in other SET schemes. One might also note that the radical anions **14** and **16** would have to be stable (and thus detectable by ESR) during the reaction, but radical anion **16** must somehow be oxidized at the end of the reaction by an unknown oxidant. Further, rather than preventing a cyclization, the mixed radical–radical anion **15**, if it could be produced, might be expected to cyclize faster than radical **8** because *intermolecular* couplings of radicals with radical anions proceed at nearly diffusion control.[41]

A simple control experiment revealed the source of the changing *E/Z* ratio for **11** observed in the Ashby studies. When an isomeric mixture of **11** was treated with excess LiAlH₄, the total amount of **11** decreased slowly from reaction with the LAH, and the isomer ratio of the remaining **11** changed with time because one isomer was destroyed faster than the other.[42] The small amount of **11**-*d₀* in the Ashby studies probably arose from the protium contamination in the sample of LiAlD₄, amplified by a normal kinetic isotope effect. The studies with iodide **13** have no bearing on those with probe **10a** because the simple tertiary radical from **13** cannot be equated with the much stabler captodative radical **9**; indeed, the Ashby group reported the detection of cyclic iodide products from **13**, confirming that this compound reacts by the radical chain isomerization pathway.

An acid test of any proposed mechanistic scheme is that the kinetics of the various steps can be used to predict the outcome of an experiment. Often one desires that the test be applied to extant results to remove the possibility of inadvertent observer bias in the data collection. Soon after Curran's group and ours reported the caveat that alkyl iodide mechanistic probe studies would be confused by the rapid atom-transfer isomerization sequence, we presented such a test in a collaborative publication.[43]

In a response to the first reports that alkyl iodide probes could provide

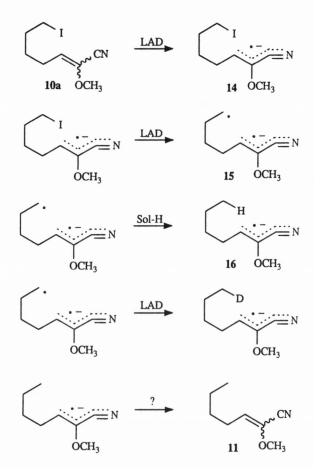

Figure 12. Mechanism proposed by the Ashby group for reactions of probe **10a**.

Substrate	Solvent	Acyclic Product	Cyclic Product
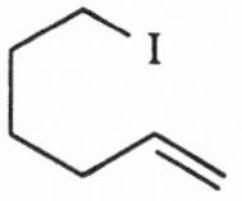 **17**	Et₂O	95% (100%-*d*)	2.5% (94%-*d*)
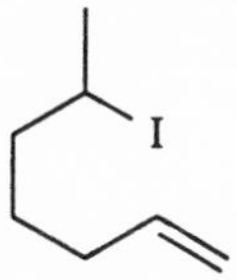 **18**	THF	35% (100%-*d*)	65% (88%-*d*)
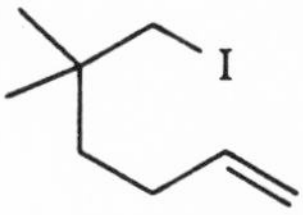 **19**	THF	5.5% (69%-*d*)	89% (59%-*d*)
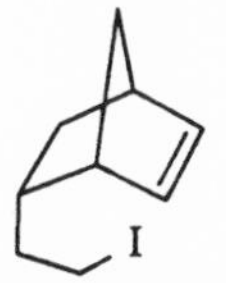 **20**	Et₂O	86% (97%-*d*)	11% (17%-*d*)

Figure 13. Results of iodide probe reductions by LiAlD₄ from ref. 10j.

misleading information, Ashby and Pham reported the results of several LiAlD₄ reductions of alkyl iodide mechanistic probes and claimed that the data supported an SET pathway in reactions of alkyl iodides with LiAlH₄.[10j] Some of their results are collected in Figure 13. Probe iodides gave acyclic and cyclic products with varying amounts of deuterium incorporation. Cyclic counterparts of three of the probes also were allowed to react with LiAlD₄. The scheme presented by Ashby's group to account for the results is given in Figure 14. It ascribes deuterated products to radical–radical reactions and undeuterated products to radical–solvent reactions. It also contains the iodine atom transfer reaction, but Ashby and Pham claimed that their results showed that even the cyclic iodide thus formed reacted by SET with the metal hydride.[10j] The scheme contains untenable radical reactions based on kinetics as well as the unusual radical cation from LiAlH₄, for which no supporting evidence exists.

The alternative scheme for reactions of alkyl iodides with LiAlD₄ is shown in Figure 15. Because most of the radical intermediates in the Ashby–Pham study were primary radicals, we had available approximate rate constants for most of the reactions. More precise rate constants for primary alkyl radical reactions at 22°C were measured[26b] and combined with known rate constants

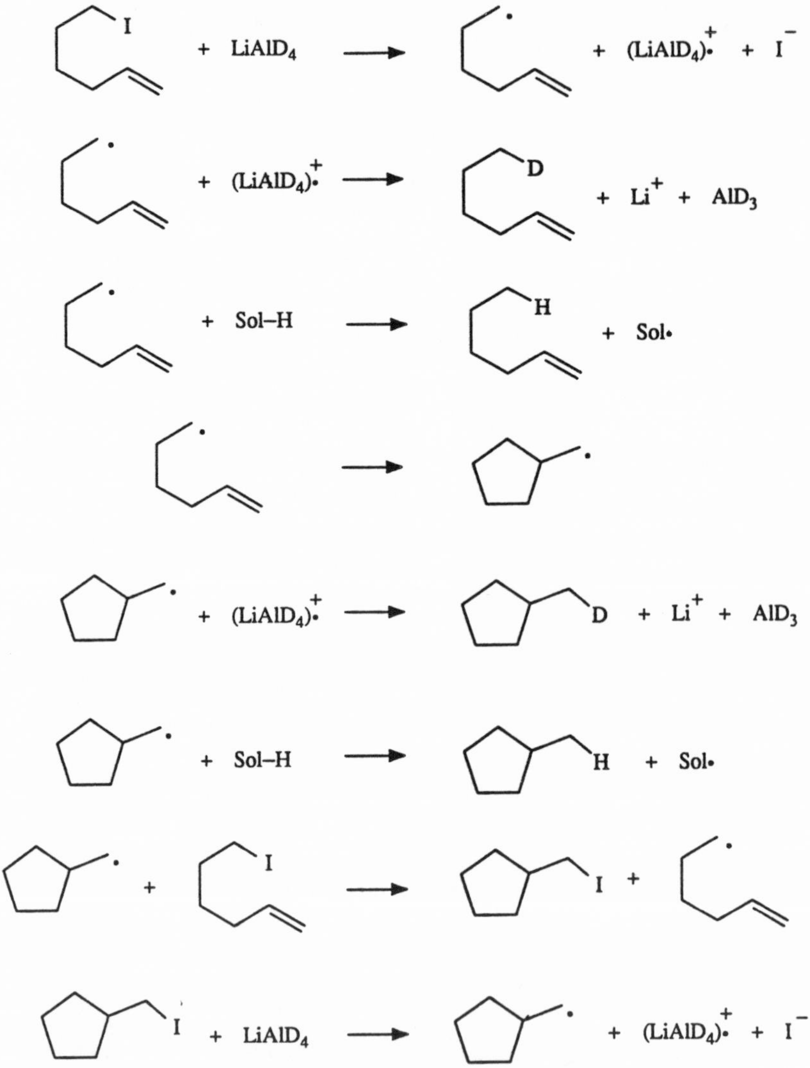

Figure 14. SET mechanism for reactions of probe iodides with LiAlD$_4$.

for radical cyclizations and a limit for the rate constant for reaction of a radical with LiAlH$_4$ to give the kinetic information necessary for analysis of the bulk of the Ashby–Pham results (Table 3).[44]

Table 4 contains the analysis. The important features are the following:

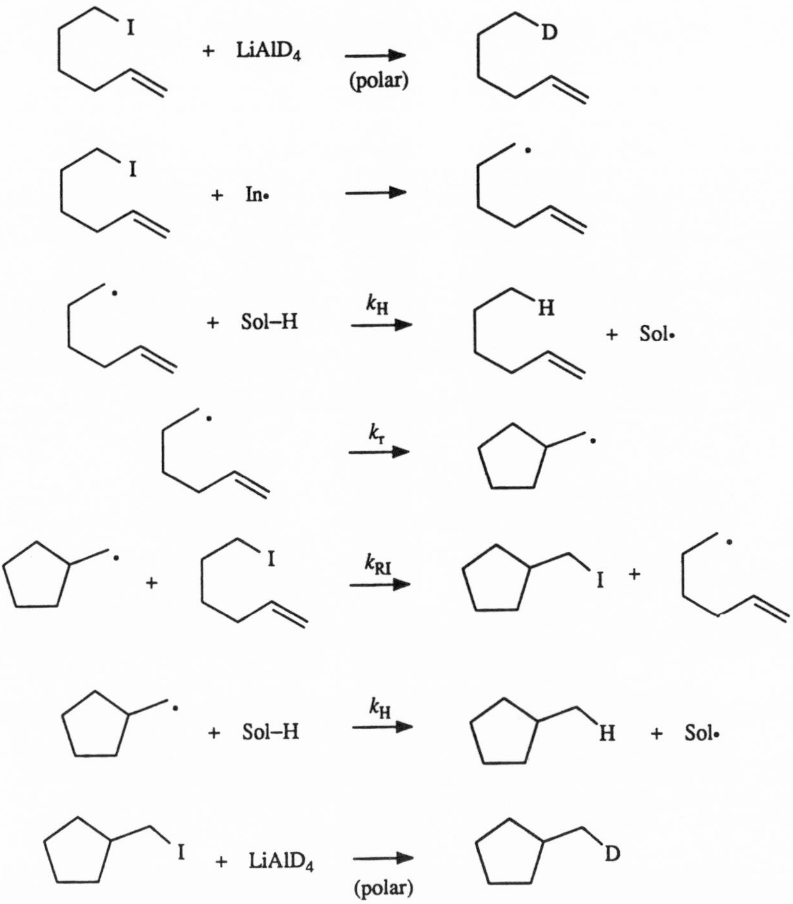

Figure 15. Polar reduction mechanism for reactions of probe iodides with LiAlD$_4$.

1. In all cases, radical cyclization was much faster than reaction of the acyclic radical with solvent or with LiAlD$_4$. Therefore, virtually all of the acyclic radicals formed must have cyclized, and the only source of acyclic material was a polar nucleophilic substitution reaction or its mechanistic equivalent in which no free intermediates were formed.
2. Based on point (1), the percentage of deuterium in the acyclic product should have been nearly the same as that in the LiAlD$_4$. In fact,

Table 3.   Rate Constants for Radical Reactions at 22°C

Reaction[a]	k	Ref.
R· + 1°–R'I → R–I + R'·	$1.2 \times 10^5 \ M^{-1} \ s^{-1}$	26b
R· + 2°–R'I → R–I + R'·	$2.6 \times 10^5 \ M^{-1} \ s^{-1}$	26b
R· + Et₂O → R–H + EtOC(·)HCH₃	$1.1 \times 10^3 \ s^{-1}$	26b
R· + THF → R–H + THF·	$6 \times 10^3 \ s^{-1}$	26b
R· + LiAlH₄ → R–H + (LiAlH₃)·	$<3 \times 10^3 \ M^{-1} \ s^{-1}$	30
5-Hexenyl cyclization	$1.9 \times 10^5 \ s^{-1}$	23
1-Methyl-5-hexenyl cyclization	$1.2 \times 10^5 \ s^{-1}$	44a
2,2-Dimethyl-5-hexenyl cyclization	$3 \times 10^6 \ s^{-1}$	44b
2-(5-*endo*-Norbornenyl)ethyl cyclization	$3 \times 10^6 \ s^{-1}$	b

[a]R· is the octyl radical.
[b]Estimated value based on the rate constant for reaction at 65°C reported in ref. 44c.

essentially complete deuterium incorporation was reported for all of the acyclic products save one that was produced in only low yield.[10j]

3. Once a cyclic radical was produced from an acyclic probe, the most important processes were reaction with the probe iodide to give the rearranged iodide and reaction with solvent to give the undeuterated cyclic product. Subsequent reaction of the cyclic iodide with LiAlD₄ by a polar process would give deuterated cyclic product. Thus, the percentage of deuterium incorporation in the cyclic product could be predicted from the rate constants for the two radical reactions (k_H and k_{RX}) and the concentrations of probe iodide. Unfortunately, concentrations were not reported in the original work; however, using typical initial probe concentrations of 0.1 and 0.2 M resulted in reasonable predictions of the amount of deuterium incorporation.

Table 4.   Analysis of Reactions of Alkyl Iodide Probes with LiAlD₄[a]

Probe[b]	Solvent	k_c/k_H[c]	Percent D in Acycle		Percent D in Cycle		
			Obs.	Calc.	Obs.	Calc.[d]	Calc.[e]
17	Et₂O	172	100	99+	94	85	92
18	THF	f	100	f	88	69	81
19	THF	500	69	98	59	50	67
20	Et₂O	3000	97	99+	17	f	f

[a]Experimental results from ref. 10j.
[b]The probes are 6-iodo-1-hexene (17), 6-iodo-1-heptene (18), 5,5-dimethyl-6-iodo-1-hexene (19), and 5-endo-(2-iodoethyl)norbornene (20).
[c]Ratio of rate constants for cyclization and reaction of the acyclic radical with solvent.
[d]Calculated with an assumed initial 0.1 M concentration of probe.
[e]Calculated with an assumed initial 0.2 M concentration of probe.
[f]Rate constants for secondary radical reactions required for these calculations.

6. RADICAL INITIATION REACTIONS IN PROBE STUDIES

The success of the analysis in Table 4 shows that we have identified the important radical reactions in an alkyl iodide mechanistic probe study. Once radical reactions were initiated, the atom transfer isomerization reaction was by far the major pathway for production of radicals from the probe. However, the radical initiation reactions were not identified nor were the chain lengths determined for the chain reactions. If the chain lengths were great, then the amount of initiation was diminishingly small, and identification of the initiation reactions is especially difficult. For example, if 5% isomerization of a probe at 0.1 M concentration was observed and the chain length was 1000, then only 0.005 mole percent initiation occurred. In this example, initiation might have resulted from impurities at 5×10^{-6} M concentration, or it might have resulted from a slow ET reaction. If one can estimate the chain length for the isomerization sequence in a mechanistic probe study, then one can extract the velocity of the radical initiation events although not necessarily the identity of these processes. Such an analysis would provide an *upper limit* on the extent of ET. In a mechanistic probe study, the probe halide is depleted by two reactions, the polar substitution reaction with velocity v_{subs} and the radical chain atom transfer isomerization reaction with velocity v_{isom}. Radical chain reactions typically proceed at relatively constant velocities throughout a reaction because steady state concentrations of the radicals are rapidly established. The velocity of the second-order polar substitution reaction will change greatly during the reaction, but, for the first half-life of the reaction of the probe, v_{subs} can be expressed as a constant, average velocity without introducing substantial errors. Typically, v_{subs} will be easier to observe, and one can relate v_{isom} to v_{subs} by the following equation:

$$v_{isom} = F \, v_{subs} \qquad (10)$$

where F is the fraction of material that isomerized.

The velocity of the isomerization process can also be expressed in terms of the elementary steps. Each propagation step in a chain sequence must have the same velocity, so the velocity for the radical chain rearrangement is given by

$$v_{isom} = k_{RX} \, [\text{R·}] \, [\text{R-X}] \qquad (11)$$

The values for k_{RX} and [R-X] in equation (11) are measurable, and v_{isom} can be measured directly or can be calculated from equation (10). Therefore, we can now calculate the concentration of radicals from equation (11). The result obtained is the average concentration of the rearranged alkyl radical during the first half-life of the reaction. However, because in the radical chain isomerization sequence the first-order rate constants for the unimolecular radical rearrangements are substantially larger than the pseudo-first-order rate constants for the bimolecular halogen atom transfer steps, the concentration of unrearranged radicals is substantially smaller than that of the rearranged radicals. This follows from the condition that the velocities of the two chain propagation steps must be the same. Thus, the calculated value for the concentration of rearranged radical is essentially equal to the total radical concentration.

As we have seen above, the radical concentration can be used to calculate the velocity of the radical termination reactions ($v_{(R\cdot R\cdot)}$). For a chain reaction sequence, the velocities of the initiation and termination reactions are equal, and we can write

$$v_{init} = v_{(R\cdot R\cdot)} = 0.25 \, k_D \, [R\cdot]^2 \qquad (12)$$

where k_D, the rate constant for diffusion at room temperature in relatively nonviscous organic solvents, is about 2×10^{10} M^{-1} s^{-1} and 0.25 is the spin statistical factor for a radical–radical reaction.[13] One can now calculate the chain length of the reaction from

$$\text{Chain length} = v_{isom}/v_{init} \qquad (13)$$

If one assumes that all radical initiation reactions involved reactions of the alkyl halide probe, one can calculate a pseudo-first-order rate constant for initiation (k'_{init}) from the velocity of the initiation process and the concentration of R-X by the following:

$$v_{init} = k'_{init} \, [R\text{-}X] \qquad (14)$$

There is a caveat. The foregoing analysis has considered rearrangements and the iodine atom transfer reactions but not reactions of radicals with solvents. For THF, hydrogen atom abstraction from solvent is about 0.1 to 0.2 times as fast as iodine atom transfer from 0.2 M alkyl iodides. If the THF radical does not react reasonably rapidly ($k > 1 \times 10^4$ M^{-1} s^{-1}) with the alkyl iodide precursor, then this radical will accumulate to concentrations higher than those estimated above for R$\cdot$. Accordingly, the termination velocity will be greater than predicted, and the chain lengths will be shorter.

Nevertheless, it is instructive to consider the analysis of some reported LiAlH$_4$ reductions of alkyl iodide probes (Table 5). Ashby's group has reported detailed studies of reactions of the probes 6-iodo-1-heptene (18),

Table 5. Analysis of LiAlH$_4$ Reactions with Iodide Probes

Probe[a]	$t_{1/2}$ (s)	$[R\cdot]$ (10^{-10} M)	Chain Length	k'_{init}[b] (10^{-8} s^{-1})	Maximum % ET[c]	Ref.[d]
18	900	37	800	70	0.1	10e
19	29,000	1.9	9300	0.19	0.01	10e
21	5400	9.3	2200	4.3	0.01	10g

[a]The probes are 6-iodo-1-heptene (**18**), 5,5-dimethyl-6-iodo-1-hexene (**19**), and 5-iodocyclooctene (**21**).
[b]Pseudo-first-order rate constant for initiation.
[c]Maximum percentage of probe that reacted by ET.
[d]Reference for experimental data.

5,5-dimethyl-6-iodo-1-hexene (**19**) and 5-iodocyclooctene (**21**), and these results are analyzed. The half-lives for isomerization in Table 5 were estimated from the reported data. The steady state concentration of radicals and chain lengths were estimated by the method previously described. Finally, the reported amount of isomerization was combined with the chain lengths to calculate the amount of radical initiation, which represents the maximum amount of ET from $LiAlH_4$ to the probe halide. The results in Table 5 admittedly are crude, but they indicate that radical initiation reactions were very minor processes in the probe studies. Studies specifically designed to determine the chain lengths of the isomerization sequences undoubtedly will lead to more precise results.

Identification of the radical initiation reactions in the probe studies probably will be the most challenging problem owing to the very small amounts of initiation that occur. Product-oriented studies involving little more than compound identifications seem to be especially poorly suited for this task, and it is likely that advances in understanding the initiation reactions will only result from a combination of kinetic studies and theoretical treatments.

Marcus theory treatments appear to be quite promising.[12] In this approach, the rate constant for an outer-sphere ET reaction is calculated from the difference between the reduction potentials of the entities and the reorganizational energy required for the ET process. Unfortunately, at this time, much of the background information for substrates of interest is not available. Furthermore, the potentials should be for reversible ET reactions, but reduction of an alkyl halide is a dissociative, irreversible event. However, if kinetic results for *unequivocal* ET reactions are determined, then one could use the theoretical approach to calculate "operational" potentials and reorganizational energies.*

A hint of the potential utility of the theoretical approach is contained in extant data. In Table 5, the calculated pseudo-first-order rate constant for initiation in the study with primary iodide probe **19** is 2×10^{-9} s^{-1}. If the initiation reaction was ET from $LiAlH_4$ (at 0.1 M) to the halide, then the second order rate constant for the ET reaction (k_{ET}) was 2×10^{-8} M^{-1} s^{-1}. In the survey of Marcus theory calculations of k_{ET} for a variety of reactions of organic substrates with

---

* The practice of estimating the electrochemical data from the kinetics of ET reactions is subject to criticism because it can result in cyclic calculations that are meaningless. Obviously, once one accepts a reaction as ET and uses the kinetics of the reaction to calculate the potentials, these potentials can only lead to a prediction that the reaction will occur by ET. At least one significant error in the estimation of an $E°$ value for a nucleophile has resulted from the approach. Eberson[12c] used the method to estimate a potential for $LiNR_2$ in the range of -1.2 V versus NHE based on purported ET reactions of $LiNR_2$ with polycyclic aromatics and benzophenone. Subsequent lithium dialkylamide probe studies with a variety of oxidants[45] and direct electrochemical measurements[46] showed that the calculated $E°$ value for $LiNR_2$ was too negative by about 1 V or 20–25 kcal/mol. Other calculated $E°$ values for nucleophiles[12c] might also be in error.

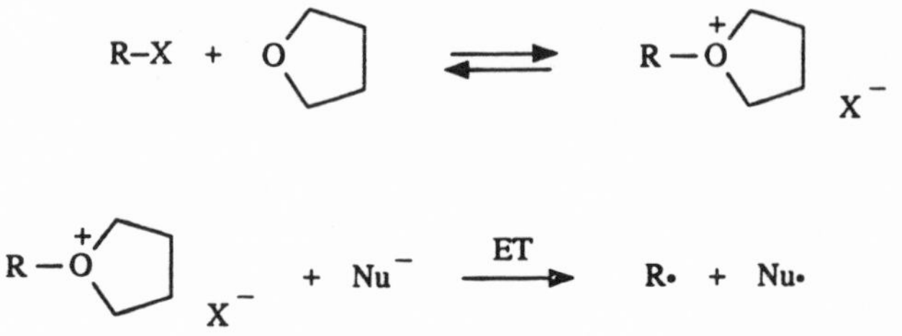

Figure 16. Possible initiation sequence in probe studies.

nucleophiles,[12c] Eberson calculated a k_{ET} for this specific reaction of 4×10^{-9} M^{-1} s^{-1}. Unlike some other cases, the data used to calculate the potential for AlH_4^- did not derive from purported SET reactions and was relatively secure.[12b] Given the crude levels of approximation both for the k'_{init} in Table 5 and for the potentials and reorganizational energies of the reaction,[12c] the agreement in the two values to a factor of five is remarkable; in fact, a concurrence within a few orders of magnitude would be considered good.

A closing speculation on the radical initiation reactions in alkyl halide studies might be in order. Based on estimated reduction potentials in the range of –0.9 to –1.1 V versus NHE for alkyl bromides and iodides,[36] a primary alkyl halide is certainly not a strong oxidant, and ET reactions with a number of relatively weak reducing agents will be slow. However, there is a polar reaction available to alkyl halides in ethereal solvents that will produce a much more powerful oxidant. Specifically, solvolysis of the alkyl halide in the ether would result in a trialkyloxonium ion. It is possible that radical initiation reactions result from such a solvolysis followed by ET from the nucleophile to the oxonium ion (Figure 16). Such a reaction pathway was found for reduction of trityl halides by lithium dialkylamide bases in THF at –78°C.[47] In THF, trityl halides produce very low concentrations of the trityl-THF oxonium in a rapid equilibrium reaction, and the powerfully oxidizing oxonium ion apparently reacts with $LiNR_2$ with a diffusion-controlled rate constant.[47] We would encourage researchers to consider the prior ionization sequence in Figure 16 as a candidate for radical initiation in future mechanistic studies of reactions of halides.

7. CONCLUSION

Despite the complications that can arise in mechanistic probe studies, the intention of this review is certainly not to dissuade investigators from using the method. In conception, the probe approach is a very powerful technique for investigating reaction pathways. The radical rearrangement, occurring with a first-order rate constant that is essentially insensitive to solvent effects, provides a precise and easy-to-use competition reaction for timing another radical reaction. The substantial amount of confusion that has resulted from studies of reactions of alkyl iodide probes with nucleophiles did not arise from an inherent inadequacy in the method but rather from inadequacies in many of the applications. Simply stated, speculations based only on the characterization of the products from a probe study have generally been superficial and of limited utility. However, when radical reaction kinetics are incorporated into the probe studies, the real power of the method is apparent. Thus, as we have seen, not only can the interfering radical chain isomerization sequence be understood, but it is also possible that the amplification of radical products owing to the chain reaction can be used to characterize slow radical initiation reactions that might involve ET processes.

This review has concentrated on studies of nucleophiles that are poor reducing agents. When good reducing agents, such as active metals, are studied, investigators can employ alkyl bromide or chloride probes in order to minimize complications owing to halogen atom transfer reactions. If one is attempting to implicate radical intermediates, then one should look for a correlation between the rate constants for radical rearrangements and the amounts of isomerized products found from a series of probes. An excellent illustration of such an application is the implication of freely diffusing radical intermediates in the formation of Grignard reagents by Garst and Whitesides.[48]

For the reactions of nucleophiles with alkyl halides, it would now appear that the most productive areas of study for probe reactions involve identification of the radical initiation reactions and accurate determinations of the chain lengths of the isomerization sequences. Toward these ends, a combination of kinetic studies and theoretical treatments is likely to prove most useful. When the chain lengths are known, extant data from probe studies can be reevaluated, and the rate constants for the radical initiation reactions (the possible ET reactions) can be calculated and compared to predictions from theory.

ACKNOWLEDGMENTS

I am grateful to my colleagues who participated in studies of radical kinetics and ET reactions: M. T. Burchill, P. W. Hurd, J. Kaplan, S.-U. Park, R. M. Sanchez, T. R. Varick, and W. G. Williams. Special thanks are due to Professor D. P. Curran for enlightening discussions and an enjoyable and productive collaboration. Kinetic studies described in this review were supported by grants from the National Science Foundation, the Robert A. Welch Foundation, and the Petroleum Research Fund administered by the American Chemical Society.

REFERENCES

1. Newcomb, M.; Smith, M. G. *J. Organomet. Chem.* **1982**, *228*, 61. Alnajjar, M. S.; Smith, G. F.; Kuivila, H. G. *J. Org. Chem.* **1984**, *49*, 1271.
2. *Chem. Eng. News* **April 13, 1981**, *26*. *Chem. Eng. News* **July 27, 1981**, *38*.

3. Ashby, E. C.; Argyropoulos, J. N. *Tetrahedron Lett.* **1984**, *25*, 7. Ashby, E. C.; Argyropoulos, J. N. *J. Org. Chem.* **1985**, *50*, 3274. Daasbjerg, K.; Lund, T.; Lund, H. *Tetrahedron Lett.* **1989**, *30*, 493.

4. Ashby, E. C.; Bae, D.-H.; Park, W.-S.; DePriest, R. N.; Su, W.-Y. *Tetrahedron Lett.* **1984**, *25*, 5107.

5. Ashby, E. C.; DePriest, R. N.; Tuncay, A.; Srivastava, S. *Tetrahedron Lett.* **1982**, *23*, 5251. Hrubiec, R. T.; Smith, M. B. *Tetrahedron* **1984**, *40*, 1457. Ashby, E. C.; Coleman, D. *J. Org. Chem.* **1987**, *52*, 4554.

6. Ashby, E. C.; Park, W. S.; Goel, A. B.; Su, W.-Y. *J. Org. Chem.* **1985**, *50*, 5184.

7. Liotta, D.; Saindane, M.; Waykole, L. *J. Am. Chem. Soc.* **1983**, *105*, 2922. Chung, S. K.; Dunn, L. B., Jr. *J. Org. Chem.* **1984**, *49*, 935. Bailey, W. G.; Gagnier, R. P.; Patricia, J. J. *J. Org. Chem.* **1984**, *49*, 2098. Bailey, W. F.; Patricia, J. J.; DelGobbo, V. C.; Jarret, R. M.; Okarma, P. J. *J. Org. Chem.* **1985**, *50*, 1999. Newcomb, M.; Williams, W. G.; Crumpacker, E. L. *Tetrahedron Lett.* **1985**, *26*, 1183. Juaristi, E.; Gordillo, B.; Aparicio, D. M.; Dulce, M.; Bailey, W. F.; Patricia, J. J. *Tetrahedron Lett.* **1985**, *26*, 1927. Ashby, E. C.; Pham, T. N.; Park, B. *Tetrahedron Lett.* **1985**, *26*, 4691. Bailey, W. F.; Patricia, J. J.; Nurmi, T. T.; Wang, W. *Tetrahedron Lett.* **1986**, *27*, 1861. Bailey, W. F.; Patricia, J. J.; Nurmi, T. T. *Tetrahedron Lett.* **1986**, *27*, 1865. Ashby, E. C.; Pham, T. N. *J. Org. Chem.* **1987**, *52*, 1291.

8. Krusic, P. J.; Fagan, P. J.; San Filippo, J., Jr. *J. Am. Chem. Soc.* **1977**, *99*, 250. San Filippo, J., Jr.; Silbermann, J.; Fagan, P. J. *J. Am. Chem. Soc.* **1978**, *100*, 4834. Newcomb, M.; Courtney, A. R. *J. Org. Chem.* **1980**, *45*, 1707. San Filippo, J., Jr.; Silbermann, J. *J. Am. Chem. Soc.* **1981**, *103*, 5588. Ashby, E. C.; DePriest, R. *J. Am. Chem. Soc.* **1982**, *104*, 6144. Kuivila, H. G.; Alnajjar, M. S. *J. Am. Chem. Soc.* **1982**, *104*, 6146. Lee, K.-W.; San Filippo, J., Jr. *Organometallics* **1983**, *2*, 906. Ashby, E. C.; DePriest, R. N.; Su, W.-Y. *Organometallics* **1984**, *3*, 1718. Alnajjar, M. S.; Kuivila, H. G. *J. Am. Chem. Soc.* **1985**, *107*, 416. Ashby, E. C.; Su, W.-Y.; Pham, T. N. *Organometallics* **1985**, *4*, 1493.

9. Ashby, E. C.; Goel, A. B.; DePriest, R. N. *J. Org. Chem.* **1981**, *46*, 2429.

10. (a) Kinney, R. J.; Jones, W. D.; Bergman, R. G. *J. Am. Chem. Soc.* **1978**, *100*, 7902. (b) Ashby, E. C.; DePriest, R. N.; Goel, A. B. *Tetrahedron Lett.* **1981**, *22*, 1763. (c) Singh, P. R.; Khurana, J. M.; Nigam, A. *Tetrahedron Lett.* **1981**, *22*, 2901. (d) Ashby, E. C.; DePriest, R. N.; Pham, T. N. *Tetrahedron Lett.* **1983**, *24*, 2825. (e) Ashby, E. C.; DePriest, R. N.; Goel, A. B.; Wenderoth, B.; Pham, T. N. *J. Org. Chem.* **1984**, *49*, 3545. (f) Ashby, E. C.; Wenderoth, B.; Pham, T. N.; Park, W.-S. *J. Org. Chem.* **1984**, *49*, 4505. (g) Ashby, E. C.; Pham, T. N. *J. Org. Chem.* **1986**, *51*, 3598. (h) Hatem, J.; Meslem, J. M.; Waegell, B. *Tetrahedron Lett.* **1986**, *27*, 3723. (i) Ashby, E. C.; Pham, T. N. *Tetrahedron Lett.* **1987**, *28*, 3183. (j) Ashby, E. C.; Pham, T. N. *Tetrahedron Lett.* **1987**, *28*, 3197. (k) Ash, C. E.; Hurd, P. W.; Darensbourg, M. Y.; Newcomb, M. *J. Am. Chem. Soc.* **1987**, *109*, 3313. (l) Pradhan, S. K.; Patil, G. S. *Tetrahedron Lett.* **1989**, *30*, 2999.

11. Pross, A. *Acc. Chem. Res.* **1985**, *18*, 212. Shaik, S. S. *Prog. Phys. Org. Chem.* **1985**, *15*, 197.

12. (a) Eberson, L. *Adv. Phys. Org. Chem.* **1982**, *18*, 79. (b) Eberson, L. *Electron Transfer Reactions in Organic Chemistry*; Springer: Berlin, 1987. (c) Eberson, L. *Acta Chem. Scand. Ser. B* **1984**, *38*, 439. (d) Lund, T.; Lund, H. *Acta Chem. Scand. Ser. B* **1986**, *40*, 470. (e) Bordwell, F. G.; Harrleson, J. A., Jr. *J. Am. Chem. Soc.* **1987**, *109*, 8112. (f) Bordwell, F. G.; Wilson, C. A. *J. Am. Chem. Soc.* **1987**, *109*, 5470.

13. Fisher, H.; Paul, H. *Acc. Chem. Res.* **1987**, *20*, 200.

14. Brace, N. O. *J. Org. Chem.* **1967**, *32*, 2711.

15. Kharasch, M. S.; Skell, P. S.; Fisher, P. *J. Am. Chem. Soc.* **1948**, *70*, 1055.

16. Russell, G. A.; Lamson, D. W. *J. Am. Chem. Soc.* **1969**, *91*, 3967. Fischer, H. *J. Phys. Chem.* **1969**, *73*, 3834. Ward, H. R.; Lawler, R. G.; Cooper, R. A. *J. Am. Chem. Soc.* **1969**, *91*, 746. Lepley, A. R.; Landau, R. L. *J. Am. Chem. Soc.* **1969**, *91*, 748.

17. Hiatt, R.; Benson, S. W. *J. Am. Chem. Soc.* **1972**, *94*, 25. Hiatt, R.; Benson, S. W. *Int. J. Chem.*

Kinet. **1972**, *4*, 151. Castelhano, A. L.; Marriott, P. R.; Griller, D. *J. Am. Chem. Soc.* **1981**, *103*, 4262.

18. For recent examples, see Curran, D. P.; Chang, C.-T. *J. Org. Chem.* **1989**, *54*, 3140. Curran, D. P.; Chen, M.-H.; Kim, D. *J. Am. Chem. Soc.* **1989**, *111*, 6265.
19. Curran, D. P.; Kim, D. *Tetrahedron Lett.* **1986**, *27*, 5821.
20. Newcomb, M.; Sanchez, R. M.; Kaplan, J. *J. Am. Chem. Soc.* **1987**, *109*, 1195.
21. Newcomb, M.; Curran, D. P. *Acc. Chem. Res.* **1988**, *21*, 206.
22. (a) Beckwith, A. L. J; Ingold, K. U. In *Rearrangements in Ground and Excited States*; deMayo, P., Ed.; Academic: New York, 1980; Vol. 1, Essay 4, pp 161–310. (b) Griller, D.; Ingold, K. U. *Acc. Chem. Res.* **1980**, *13*, 317.
23. Chatgilialoglu, C.; Ingold, K. U.; Scaiano, J. C. *J. Am. Chem. Soc.* **1981**, *103*, 7739.
24. Barton, D. H. R.; Crich, D.; Motherwell, W. B. *Tetrahedron* **1985**, *41*, 3901.
25. Newcomb, M.; Kaplan, J. *Tetrahedron Lett.* **1987**, *28*, 1615.
26. (a) Newcomb, M.; Park, S. U. *J. Am. Chem. Soc.* **1986**, *108*, 4132. (b) Newcomb, M.; Kaplan, J. *Tetrahedron Lett.* **1988**, *29*, 3449.
27. Hawari, J. A.; Engel, P. S.; Griller, D. *Int. J. Chem. Kinet.* **1985**, *17*, 1215.
28. Johnston, L. J.; Lusztyk, J.; Wayner, D. D. M.; Abeywickreyma, A. N.; Beckwith, A. L. J.; Scaiano, J. C.; Ingold, K. U. *J. Am. Chem. Soc.* **1985**, *107*, 4594.
29. Russell, G. A.; Guo, D. *Tetrahedron Lett.* **1984**, *25*, 5239.
30. Beckwith, A. L. J.; Goh, S. H. *J. Chem. Soc., Chem. Commun.* **1983**, 907.
31. Giles, J. R. M., Roberts, B. P. *J. Chem. Soc., Chem. Commun.* **1981**, 1167.
32. Franz, J. A.; Barrows, R. D.; Camaioni, D. M. *J. Am. Chem. Soc.* **1984**, *106*, 3964.
33. Russell, G. A.; Danen, W. C. *J. Am. Chem. Soc.* **1968**, *90*, 347. Kornblum, N. *Angew. Chem., Int. Ed. Engl.* **1975**, *14*, 734. Bunnett, J. F. *Acc. Chem. Res.* **1978**, *11*, 413.
34. (a) Russell, G. A.; Ros, F.; Mudryk, B. *J. Am. Chem. Soc.* **1980**, *102*, 7601. (b) Russell, G. A.; Hershberger, J.; Owens, K. *J. Organomet. Chem.* **1982**, *225*, 43. (c) Russell, G. A.; Khanna, R. K. *Tetrahedron* **1985**, *41*, 4133.
35. Andrieux, C. P.; Iluminada, G.; Savéant, J.-M. *J. Am. Chem. Soc.* **1989**, *111*, 1620.
36. Ref. 12b, p 52.
37. Sim, B. A.; Griller, D.; Wayner, D. D. M. *J. Am. Chem. Soc.* **1989**, *111*, 754.
38. (a) Park, S.-U.; Chung, S. K.; Newcomb, M. *J. Am. Chem. Soc.* **1986**, *108*, 240. (b) Viehe, H. G.; Janousek, Z.; Merenyi, R.; Stella, L. *Acc. Chem. Res.* **1985**, *18*, 148.
39. Park, S.-U.; Chung, S.-K.; Newcomb, M. *J. Org. Chem.* **1987**, *52*, 3275.
40. (a) Ashby, E. C. *Acc. Chem. Res.* **1988**, *21*, 414. (b) Ashby, E. C.; Pham, T.; Madjdabadi, A. A. *J. Org. Chem.* **1988**, *53*, 6156.
41. Hebert, E.; Mazaleyrat, J. P.; Welvart, Z.; Nadjo, L.; Savéant, J.-M. *Nouv. J. Chim.* **1985**, *9*, 75.
42. Varick, T. R.; Newcomb, M., unpublished results, Texas A&M University.
43. Newcomb, M.; Kaplan, J.; Curran, D. P. *Tetrahedron Lett.* **1988**, *29*, 3451.
44. (a) Lusztyk, J.; Maillard, B.; Deycard, S.; Lindsay, D. A.; Ingold, K. U. *J. Org. Chem.* **1987**, *52*, 3509. (b) Beckwith, A. L. J.; Easton, C. J.; Lawrence, T.; Serelis, A. K. *Aust. J. Chem.* **1987**, *36*, 545. (c) Ashby, E. C.; Pham, T. N. *Tetrahedron Lett.* **1984**, *25*, 4333.
45. Newcomb, M.; Burchill, M. T.; Deeb, T. M. *J. Am. Chem. Soc.* **1988**, *110*, 6528.
46. Renaud, P.; Fox, M. A. *J. Am. Chem. Soc.* **1988**, *110*, 5702.
47. Newcomb, M.; Varick, T. R.; Goh, S.-H., *J. Am. Chem. Soc.* **1990**, *112*, 5186.
48. Garst, J. F.; Deutch, J. E.; Whitesides, G. M. *J. Am. Chem. Soc.* **1986**, *108*, 2490. Garst, J. F.; Swift, B. L.; Smith, D. W. *J. Am. Chem. Soc.* **1989**, *111*, 234. Garst, J. F.; Swift, B. L. *J. Am. Chem. Soc.* **1989**, *111*, 241.

FREE RADICAL REACTIONS

FRAGMENTATION AND REARRANGEMENTS IN

AQUEOUS SOLUTION

Michael J. Davies and Bruce C. Gilbert

Advances in Detailed Reaction Mechanisms
Volume 1, pages 35–81
Copyright © 1991 JAI Press Inc.
All rights of reproduction in any form reserved.
ISBN: 1-55938-164-7

1. INTRODUCTION

The continued and widespread interest in the generation and reactions of free radicals reflects a number of factors, including the recognition that their unique properties may be harnessed in sophisticated yet simple approaches to synthetic strategies,[1] the relevance of free-radical chemistry to combustion and autoxidation processes,[2] and the belief that free radicals may play a significant role in biological systems.[3] For example, it has been claimed that free radicals are involved in photosynthesis and respiration, in the action of and response to certain xenobiotics (e.g., CCl_4), and in the critical stages of radiation damage.[3]

The rich variety and types of reaction that free radicals can undergo—-including, for example, hydrogen atom abstraction, addition to unsaturated systems (e.g., alkenes and arenes), rearrangements and fragmentation, termination (dimerization and disproportionation), and electron transfer reactions (both oxidation and reduction)—have been widely reviewed[2] and most of these reactions will not be discussed further here. Instead, attention will be focused on some of the unusual reactions that are encouraged when radicals are generated in aqueous solution—many examples of which are susceptible to acid and/or base catalysis and some of which may involve novel short-lived radical cations or radical anions. This review, which is intended to complement to some extent the more detailed surveys of radical rearrangements (largely in nonpolar solvents),[4] will illustrate how radical behavior is influenced by a polar solvent (and the possibilities for protonation and deprotonation), and show how the mechanistic information obtained is of considerable assistance in building up an understanding of radiation damage in biological systems. Particular emphasis will therefore be placed on the behavior of free radicals derived from 1,2-diols [e.g., $\cdot CH(OH)CH_2OH$] and related compounds (as the basis for the discussion of the more complex behavior of carbohydrates and nucleic acids (see, e.g., refs. 5 and 6). It will also be shown how related radicals may be derived by addition reactions (e.g., of $\cdot OH$ to arenes, including aromatic amino acids and alkenes, and to the sulfur substituent in, for example, methionine).

Most of the research that will be described has involved both the generation of free radicals (by a variety of techniques) and the mechanistic study (by spectroscopic and kinetic methods) of the first-formed short-lived intermediates. Two techniques are highlighted here. The use of *pulse radiolysis*[7] of aqueous solutions allows the hydroxyl radical to be readily generated (along with e^- and, to a much smaller extent, $H\cdot$); addition of scavengers for the hydrated electron allows the reactions of the hydrogen atom and, especially, of the hydroxyl radical to be explored [see reactions (1)–(4)]. In many of these experiments the use of spectrophotometric detection allows the ultraviolet (UV)-visible absorption spectrum of an intermediate radical to be monitored (as a function of λ) and its decay followed within a short period (typically microseconds) of the pulse. Radical yields and products can be determined in

$$H_2O \xrightarrow{\gamma} e^- + H^. + HO^. \tag{1}$$

$$e^- + N_2O \xrightarrow{H_2O} HO^. + N_2 + HO^- \tag{2}$$

$$e^- + H^+ \longrightarrow H^. \tag{3}$$

$$HO^. + RH \longrightarrow R^. + H_2O \tag{4}$$

suitable experiments, and conductivity detection offers an attractive alternative, suitable for the characterization of reactions involving ions (see Section 4).

Electron spin resonance (ESR) spectroscopy offers what is in many ways a complementary technique for the direct detection and structural characterization of free radicals (present at concentrations greater than circa 10^{-7} mol dm^{-3}).[8-10] Experiments are normally carried out under steady state conditions (whereby radicals are continuously generated), so that the full ESR spectrum, rich in hyperfine splittings (and structural detail), can be obtained via a magnetic field scan [as in a continuous wave (CW) NMR experiment]. Methods employed to generate free radicals *continuously* include radiolysis (as mentioned previously), direct photolysis [e.g., with a laser or UV lamp) of solutions of suitable substrates [e.g., hydrogen peroxide, reaction (5), or other peroxides] or the rapid reaction of certain redox couples [e.g., reaction (6) or the so-called Fenton reaction (7)], brought about in the cavity of the spectrometer via the use of continuous mixing via rapid-flow systems. Addition of, for example, ethane-1,2-diol in such a system leads to the detection of the ESR signal of $\cdot CH(OH)CH_2OH$ at pH 7; as the pH is lowered, its signal is accompanied by and replaced by that from $\cdot CH_2CHO$[11] [see reactions (8) and (9) and Figure 1]. Though this is not a time-resolved experiment, kinetic information about the transformation can be obtained, as described in Section 2.

$$H_2O_2 \xrightarrow{h\upsilon} 2\,HO^. \tag{5}$$

$$Ti^{III} + H_2O_2 \longrightarrow Ti^{IV} + HO^. + HO^- \tag{6}$$

$$Fe^{II} + H_2O_2 \longrightarrow Fe^{III} + HO^. + HO^- \tag{7}$$

$$HO^. + CH_2(OH)CH_2OH \longrightarrow H_2O + \cdot CH(OH)CH_2OH \tag{8}$$

$$\cdot CH(OH)CH_2OH \xrightarrow[-H^+]{+H^+} \cdot CH_2CHO + H_2O \tag{9}$$

It is our intention here to show how radiolytic techniques and redox couples have been used to unravel the mechanism of reaction (9) and many analogous reactions involving species of the type $\cdot CHXCH_2Y$. As will be seen, most of them have in common the presence of an electron-donating (+M) substituent (X) at the α carbon (radical center) and, at the β carbon, a –I substituent Y or, put another way, a good anionic leaving group. Other

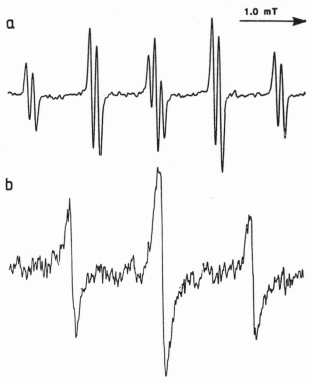

Figure 1. ESR spectra of ·CH(OH)CH$_2$OH (a) and ·CH$_2$CHO (b) obtained from reaction of ·OH (from Ti^{III}/H$_2$O$_2$) and ethane-1,2-diol at pH values of 4 and 1, respectively.

examples that will be discussed have Y = Cl, OAc, OMe, NH$_3$$^+$, and phosphate, for some of which the state of protonation (i.e., as governed by the pH) may be important. We will also describe the investigation of the role of the α substituent (e.g., OH, OR, O$^-$) and explore the possibility that under suitable conditions even ordinary β-oxygen-substituted radicals (e.g., ·CH$_2$CMe$_2$OH or ·CH$_2$CMe$_2$OAc) can undergo ready rearrangement. In even the simplest reaction [reaction (9)] it appears that loss of H$^+$ (from α OH) and OH$^-$ is involved, and these steps could of course be consecutive or synchronous: in the event that OH$^-$ is lost first (as H$_2$O, in an acid-catalyzed process) a novel radical cation HOĊH-ĊH$_2$ would be formed. Several parts of this review are concerned with experiments to prove or disprove the involvement of such species, and, by a variety of other approaches, to explore the properties of these novel intermediates.

2. DEHYDRATION OF RADICALS FROM 1,2-DIOLS: A KINETIC STUDY OF SOME SIMPLE EXAMPLES

Evidence for the acid-catalyzed conversion of radicals $\cdot CR^1(OH)CR^2R^3OH$ into $\cdot CR^2R^3C(O)R^1$ [reaction (10)] and values for the rate constants of reaction can be obtained by following the first-order buildup in the UV absorption from the carbonyl-conjugated radical at circa 240 nm, as the α-hydroxyalkyl radical is replaced in a few microseconds after the pulse in a pulse–radiolysis experiment (see Figures 2 and 3).[12,13]

$$\cdot CR^1(OH)CR^2R^3OH + H^+ \quad \xrightarrow[-H_2O]{k_{10}} \quad R^1C(O)\dot{C}R^2R^3 \qquad (10)$$
$$\quad S_1 \qquad\qquad\qquad\qquad\qquad\qquad S_2$$

The rate of the buildup (and hence the pseudo-first-order rate constant k_{obs}) depends on pH, and knowledge of $[H^+]$ allows the second-order (overall) rate constant (k_{10}) to be determined (see, e.g., Figure 4), though the separate behavior of isomeric radicals in a mixture cannot easily be distinguished.

In parallel ESR experiments[13] the appearance of the carbonyl-conjugated radical (recognized with its different splittings and g value) can be followed as a function of pH in flow experiments (and mixtures of isomeric carbonyl or hydroxyalkyl radicals can normally be distinguished). It can be shown by steady

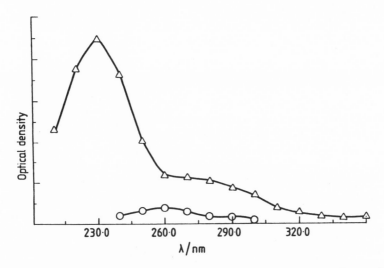

Figure 2. Absorptions detected during pulse–radiolysis of solutions of 2-methylpropane-1,2-diol (0.1 mol dm^{-3}) saturated with N_2O at pH 6: $\circ$, 2–6 μs after pulse, $\cdot CH(OH)CMe_2OH$; $\triangle$, 60–80 μs after pulse, $\cdot CMe_2CHO$.

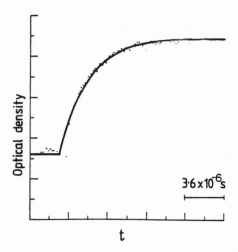

Figure 3. Buildup of the absorbance from $\cdot CMe_2CHO$ at 240 nm in a pulse--radiolysis experiment on $CH_2(OH)CMe_2OH$ at pH 3 (see Figure 2).

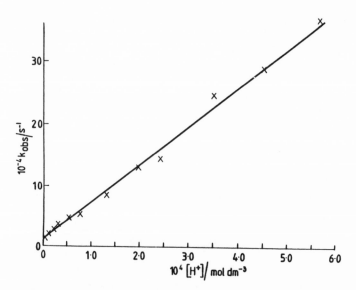

Figure 4. Variation of k_{obs} for buildup of absorption from $\cdot CMe_2CHO$ [from $\cdot CH(OH)CMe_2OH$] with [H$^+$] (see Figure 3).

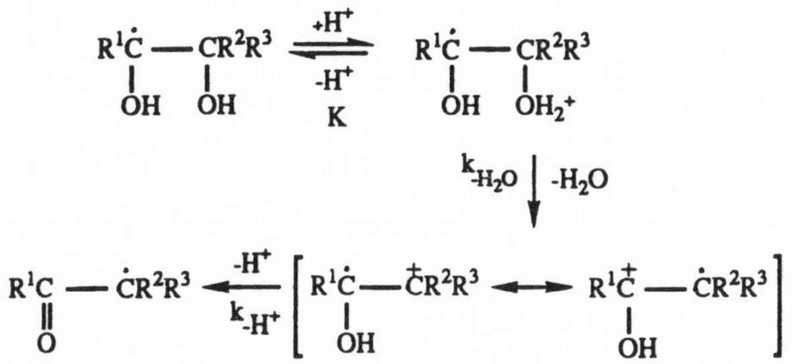

$$\begin{array}{c}
\underset{\substack{|\\ OH}}{R^1\overset{\cdot}{C}} - \underset{\substack{|\\ OH}}{CR^2R^3} \underset{\underset{K}{-H^+}}{\overset{+H^+}{\rightleftharpoons}} \underset{\substack{|\\ OH}}{R^1\overset{\cdot}{C}} - \underset{\substack{|\\ OH_2^+}}{CR^2R^3}
\end{array}$$

$$k_{-H_2O} \downarrow -H_2O$$

$$\underset{\substack{\|\\ O}}{R^1C} - \overset{\cdot}{C}R^2R^3 \underset{\underset{k_{-H^+}}{-H^+}}{\longleftarrow} \left[\underset{\substack{|\\ OH}}{R^1\overset{\cdot}{C}} - \overset{+}{C}R^2R^3 \longleftrightarrow \underset{\substack{|\\ OH}}{R^1\overset{+}{C}} - \overset{\cdot}{C}R^2R^3 \right]$$

Scheme 1

state analysis that if ·OH [reaction (6)] is scavenged by the substrate (SH) to give a first-formed radical S_1· and if S_1· disappears both by radical–radical recombination (with rate constant k_t) and, in acid, by rearrangement to give S_2· (which also decays by diffusion-controlled self- and cross-termination reactions), then

$$k_6[Ti^{III}]_t[H_2O_2]_t = k_{10}[S_1·][H^+] + k_t[S_1·]([S_1·] + [S_2·]) \qquad (11)$$

Plots of $1/[S_1·]$ against $[H^+]$ allow k_{10} (and k_t) to be determined.

The detailed data obtained using these approaches (see Table 1) have been used both to test the applicability of the mechanism proposed by Henglein and co-workers[12] to account for the rearrangement (see, e.g., Scheme 1) and to gain a further insight into the factors that control the reaction.

As can be seen, it is proposed that a *radical cation* is formed after protonation and loss of the β OH group. (Intuitively this is in accord with the observation that the rate of reaction is accelerated by the presence of alkyl groups: compare the overall rate constants of 2.2×10^7 and 1.5×10^8 dm^3 mol^{-1} s^{-1} for ethane-1,2-diol and butane-2,3-diol, respectively.) More detailed kinetic analysis of the pulse–radiolysis experiments indicates that if the equilibrium for the first step is rapidly attained and if deprotonation of the radical cation is faster than loss of water, then the half-life for the build up of S_2· is given by the relationship

$$\tau_{1/2} = \frac{\ln 2}{k_{-H_2O}} + \frac{K \ln 2}{k_{-H_2O} [H^+]} \qquad (12)$$

where $K = [H^+][·CR^1(OH)CR^2R^3OH]/[·CR^1(OH)CR^2R^3(OH_2^+)]$. This is shown plotted in Figure 5, from which it can be seen that the predicted behavior is followed and that k_{-H_2O} and K (see Scheme 1) can be determined.

Table 1. Rate Constants for the Acid-Catalyzed Dehydration of Radicals from 1,2-Diols and Related Compounds[a]

Compound	$k_{10}/dm^3\,mol^{-1}\,s^{-1}$ (ESR)[b]	$k_{10}/dm^3\,mol^{-1}\,s^{-1}$ (PR)[c]	k_{-H_2O}/s^{-1}	$K/mol\,dm^{-3}$
Ethane-1,2-diol	2.1×10^7	1.1×10^7	7.5×10^5	4.6×10^{-2}
Propane-1,2-diol	$\left\{\begin{array}{l}4.1 \times 10^{8\ d}\\ 1.8 \times 10^{8\ e}\end{array}\right.$	9.0×10^7	7.1×10^5	5.6×10^{-3}
Butane-1,2-diol		1.5×10^8	8.5×10^5	5.3×10^{-3}
2-Methylpropane-1,2-diol		7.0×10^8	2.7×10^6	3.7×10^{-3}
Butane-2,3-diol	1.0×10^9	9.7×10^8	2.8×10^6	2.6×10^{-3}
2-Methylbutane-2,3-diol		1.7×10^9	1.8×10^6	7.4×10^{-4}
Cyclohexane-1,2-diol		1.3×10^9	1.8×10^6	1.5×10^{-3}
Glycerol	$\left\{\begin{array}{l}1.0 \times 10^{7\ f}\\ 4.6 \times 10^{6\ g}\end{array}\right.$	3.5×10^6	6.5×10^5	1.3×10^{-1}
Erythritol	6.0×10^6	2.9×10^6	3.5×10^5	8.0×10^{-2}
Myoinositol	$\left\{\begin{array}{l}7.0 \times 10^{5\ h}\\ 5.7 \times 10^{4\ i}\end{array}\right.$	5.7×10^5	2.4×10^5	3.2×10^{-1}
2-Methoxyethanol[j]		1.0×10^5	4.8×10^5	2.35
2-Hydroxymethyl-tetrahydrofuran[k]		4.9×10^5	1.0×10^5	4.7×10^{-2}
2-Deoxyribose		2.8×10^6	4.8×10^5	8.4×10^{-2}
		1.5×10^7	7.6×10^5	4.0×10^{-2}

[a]For definitions of k_{-H_2O} and K, see Scheme 1. [b]±30%. [c]±10%. [d]$CMe(OH)CH_2OH$. [e]$\dot{C}H(OH)CHMeOH$. [f]$\dot{C}Me(OH)CH_2OH$. [g]$\dot{C}H(OH)CH(OH)$-CH_2OH. [h]Isomer with β OH equatorial. [i]Isomer with β OH axial. [j]$\dot{C}H(OH)CH_2OMe$ to $\cdot CH_2CHO$. [k]For reaction of $OCH_2CH_2CH_2\dot{C}HCHOH$ to $HOCH_2$-$CH_2CH_2\dot{C}HCHO$.

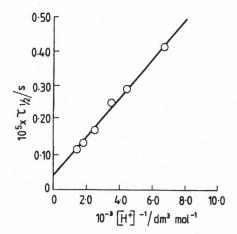

Figure 5. Variation with $[H^+]^{-1}$ of $\tau_{1/2}$ for the buildup of the cyclohexanone-2-yl radical following reaction of cyclohexane-1,2-diol with ·OH (generated by pulse radiolysis).

For the range of substrates studied, the values of k_{-H_2O}, the rate constant for the overall loss of water from the protonated radical, show that electron-withdrawing groups (e.g., hydroxymethyl) slow down the dissociation, as would be expected if a radical cation is indeed formed. Parallel changes are found in the magnitudes of K values. The structural dependence of the pK_a values is in the direction predicted (i.e., +I substituents decrease the acid strength, –I substituents cause a significant increase), but the absolute magnitudes of pK (up to 3.1) are generally higher than might have been expected from the pK value for the parent compounds (≤ 0.34). This may reflect the electron-releasing +M effect of the α oxygen (which should make it easier for this to accept a positive charge) as well as the anticipated favorable interaction between the unpaired electron and the σ bond to the β OH_2^+ group. [It is known[14] that an eclipsed geometry (**1**) is favored by α,β-dihydroxyalkyl radicals as a result of the combination of +M (α) and –I (β) effects (cf. canonical structures **3** and **4**). The high value of K (2.35), and hence low basicity, for the isomeric radical from myoinositol with *equatorial* β OH groups (evidence for which argument is described subsequently) supports this proposal.

The ESR experiments, which possess extra resolution in that mixtures of radicals with different splittings can be readily analyzed, allow us to distinguish, for example, the different behavior of ·CMe(OH)CH$_2$OH and ·CH(OH)CHMeOH (from propane-1,2-diol) and also establish that ·C(OH)(CH$_2$OH)CH(OH)CH$_2$OH, from erythritol, gives predominantly ·CH$_2$C(O)CH(OH)CH$_2$OH rather than ·CH(CH$_2$OH)C(O)CH$_2$OH.[6]

It has also been demonstrated[12,15] that base-catalyzed dehydration of 1,2-dihydroxyalkyl radicals can occur, the initial step being ionization of the α hydroxyl group (see Scheme 2). The rate constants for steps (i) and (ii) have been shown by pulse radiolysis studies to be 10^{10} dm^3 mol^{-1} s^{-1} and $\geq 10^5$ s^{-1}, respectively; in a subsequent study of a number of alkylated analogs it was shown[13] that the rate constants for the loss of hydroxide ion range from 2.6×10^6 s^{-1} [for ·C(CH$_2$OH)(O$^-$)CH(CH$_2$OH)OH] and 3.1×10^6 s^{-1} [for ·CH(O$^-$)-CH$_2$OH] to $\geq 8.1 \times 10^6$ s^{-1} [for ·CMe(O$^-$)CHMeOH], a result that is interpreted in terms of the stabilization of the transition state by hyperconjugation involving alkyl groups.

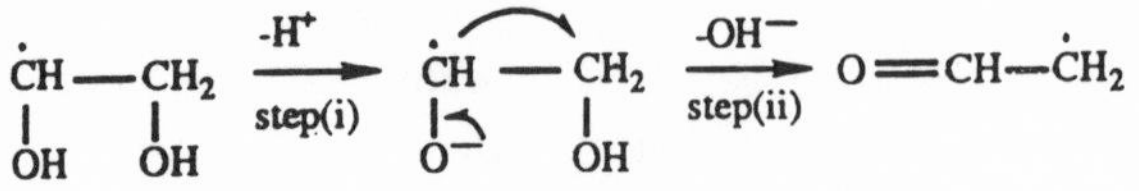

$$\overset{\cdot}{C}H-CH_2 \quad \xrightarrow[step(i)]{-H^+} \quad \overset{\cdot}{C}H - CH_2 \quad \xrightarrow[step(ii)]{-OH^-} \quad O = CH-\overset{\cdot}{C}H_2$$
$$\underset{OH}{|} \quad \underset{OH}{|} \qquad\qquad \underset{O^-}{|} \quad \underset{OH}{|}$$

Scheme 2

The observation that for some radicals (e.g., that from 2-methylbutane-1,2-diol) rapid reaction gives the carbonyl-conjugated radical even under nearly *neutral* conditions indicates that an uncatalyzed or *spontaneous* loss of water can occur; this is also indicated by the observation of an intercept in Figure 4 and by the fact that rearranged radicals can be observed directly by ESR in aqueous solution (see Figure 6, which also reveals the importance of the polar

$$·CMe(OH)CMe_2OH \quad \xrightarrow[pH \ ca. \ 6.5]{k \ 2 \times 10^4 \ s^{-1}} \quad H^+ + ·CMe_2C(O)Me + HO^- \qquad (13)$$

solvent in encouraging fragmentation). More detailed studies show that k for the spontaneous loss of water [reaction (13)] is 2×10^4 s^{-1} at 20°C (with E_a 19 kJ mol^{-1} and $\Delta S^{\ddagger}$ –93 J mol^{-1} K^{-1}). It is believed that the presence of the two β methyl groups helps lock the radical in the eclipsing conformation (1), and hence increase overlap between the β C–O bond and the radical's singly occupied molecular orbital (SOMO), and that deprotonation of α OH (which

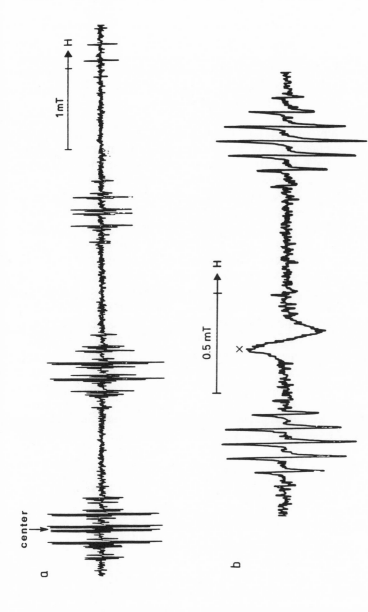

Figure 6. (a) ESR spectrum (half) of ·CMe₂C(O)Me obtained from photolysis of an *aqueous* solution of CHMe(OH)CMe₂OH, propanone, and potassium peroxydisulfate (pH range 3–6, circa 4°C). (b) ESR spectrum of ·CH(OH)CMe₂OH obtained by photolysis of CH₂(OH)CMe₂OH and di-*tert*-butyl peroxide in propanone at circa 4°C. The peak marked (×) is from ·CH₂COMe, derived from the solvent.

45

leads to the negative entropy change via solvation of H^+) is concerted with loss of OH^- (aided by the steric and electronic effects of the methyl groups).

More detailed steric and electronic probes for this type of reaction are provided by sugars and related molecules, discussed in the next section.

3. FRAGMENTATION REACTIONS OF α,β-DIOXYGEN-SUBSTITUTED RADICALS FROM CARBOHYDRATES

The radiation chemistry of aqueous solutions of carbohydrates and polynucleotide components has been studied extensively, mainly via product studies following attack by ·OH (generated by radiolysis) on model compounds.[5,6] It has been suggested that radiation damage—induced by ·OH—involves a variety of possible secondary reactions of first-formed radicals, some of which are closely related to those described previously. For instance, DNA strand breakage is believed to involve heterolytic fragmentation of sugar-derived radicals, these having been derived largely via secondary reactions of base-derived species (examples will be discussed subsequently). Although ESR spectroscopy is usually unable to provide the kind of detailed *kinetic* analysis already described (and obtained by pulse–radiolysis in relatively simple systems), its use in conjunction with metal–ion redox couples does allow direct identification of complex mixtures of radicals formed in the reaction of ·OH and carbohydrates; their fragmentation pathways can also often be deduced.

For example, ESR spectroscopy has been employed[16] to show that reaction of ·OH with the model substrate myoinositol (**5**) is unselective: at pH circa 4 signals of all possible radicals produced by C–H abstraction can be detected and identified on the basis of their β proton splitting constants (note that a large splitting of circa 3.0 mT characterizes an *axial* β H, with small dihedral angle between the C–H bond and the unpaired electron's orbital). As the pH is

(5)   (6)   (7)

lowered, these radicals are transformed into carbonyl-conjugated radicals (by a mechanism as described earlier): the radical with an axial β hydroxyl group (**6**) is found to lose water (to give **7**) with a rate constant ($3 \times 10^6 \, dm^3 \, mol^{-1} \, s^{-1}$) significantly greater than those of radicals with only equatorial β OH groups, indicative of a stereoselective requirement for loss of $-OH_2^+$ (involving overlap between the β C–O bond and the SOMO).

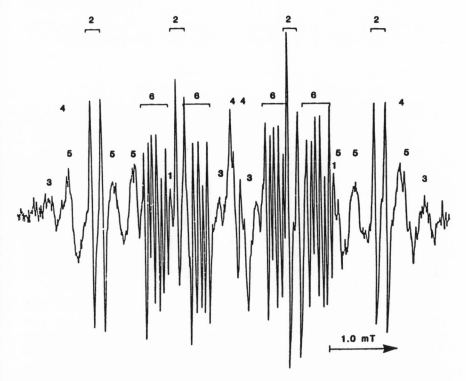

Figure 7. ESR spectra of hydrogen-abstraction radicals derived by reaction of α-D-glucose with ·OH (from Ti^{III} and H$_2$O$_2$) in aqueous solution at pH circa 4.

Again, for both anomers of glucose all six radicals formed via (relatively unselective) C–H abstraction can be identified at pH circa 4 and a full conformational analysis obtained (see Figure 7 and ref. 17). As the pH is lowered the signals from all six α,β-dioxygen-substituted radicals are removed, as rearrangements occur akin to those described previously. The fastest rearrangement for α-D-glucose (and hence the first to be revealed as the pH is lowered, to circa 2.8) is that of radical **8** to **9** [reaction (14), for which a rate constant of 5×10^6 dm^3 mol^{-1} s^{-1} is estimated]: this process is believed to be enhanced by the production of a radical stabilized by both +M and −M groups and by the favorable geometry for hydroxyl loss. The C_1 radical (10) rearranges to give 11 [reaction (15)].

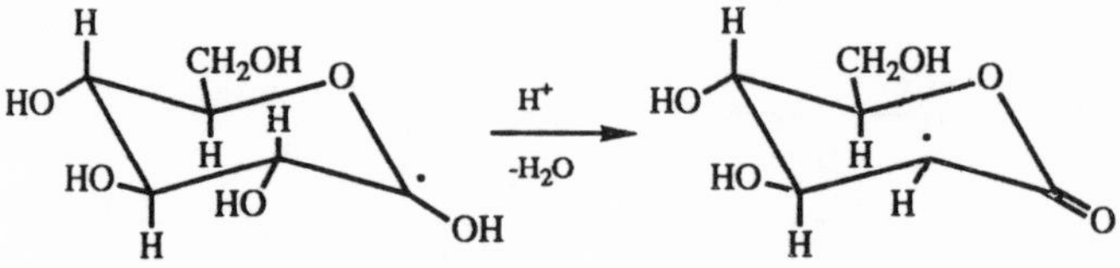

(15)

(10)   (11)

Evidence has also been presented that radicals with α or β alkoxyl groups (e.g., those derived from C_5 and C_6, respectively) can rearrange via related processes.[17] As described below it is believed that the appropriate C_5 radical can lose a β OH group (but not of course a proton from the α group) to give a radical cation [see reaction (16)]. The C_6 radical effectively undergoes acid-catalyzed loss of OR^- via ring scission [see reaction (17), which has a k of 10^5 dm^3 mol^{-1} s^{-1}] and reaction (18) shows the analogous acid-catalyzed fragmentation of a radical from 1-O-methyl-β-D-glucose (the relatively high value of k, 3.5×10^5 dm^3 mol^{-1} s^{-1}, probably reflects the axial disposition of the leaving group). Reactions analogous to (18), and catalyzed by acid or base, provide a plausible mechanism for radical induced degradation of polysaccharides (e.g., dextran: see ref. 18).

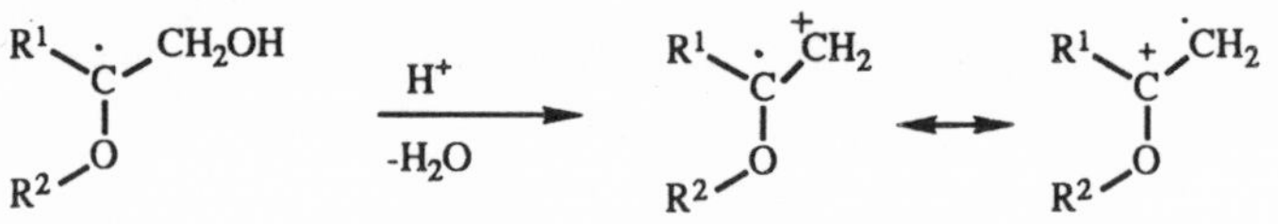

(16)

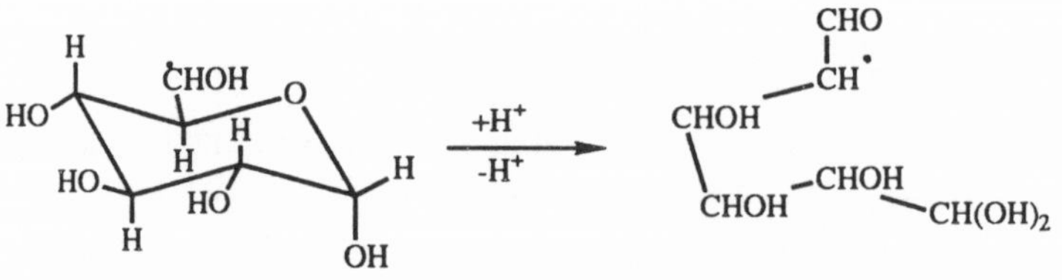

(17)

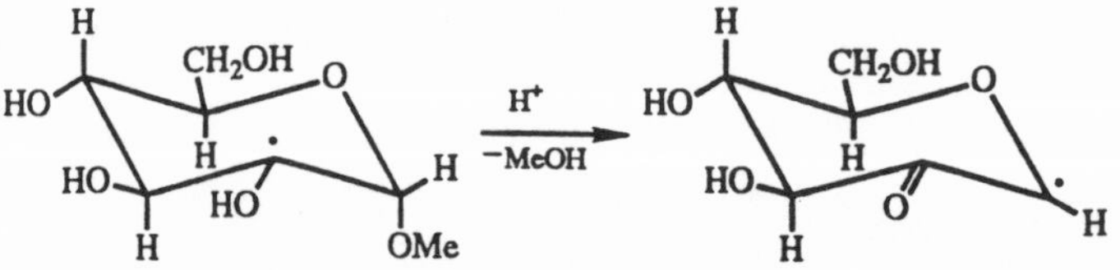

(18)

Reactions of this type have been demonstrated for a wide range of other carbohydrates, including aldopentoses, sucrose, and other compounds contain-

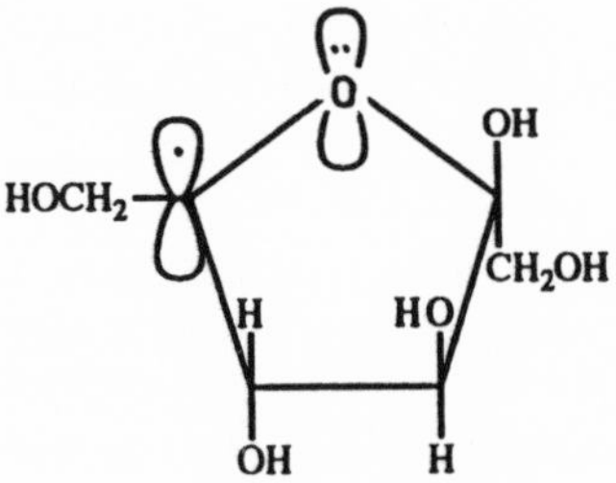

(12)

ing furanose rings, including models for DNA.[19] One of the most interesting
findings is that, whereas in pyranose compounds radicals appear to be formed
in relatively unselective attack, reaction of ·OH with sucrose and β-D-fructofuranose occurs preferentially at position C_{5-H} in the furanose ring(to give **12**
from the latter): this is believed to reflect the operation of a *stereoelectronic
effect* whereby a favorable geometry allows a stabilizing interaction between
the lone pair of electrons on oxygen and the developing radical center.

As noted earlier, base-catalyzed analogs of these reactions are believed to
proceed via the appropriate ionized species [i.e., the ketyl radicals ·CR(O⁻)
CR_2OH] in which the +M effect of the α substituent is thereby considerably

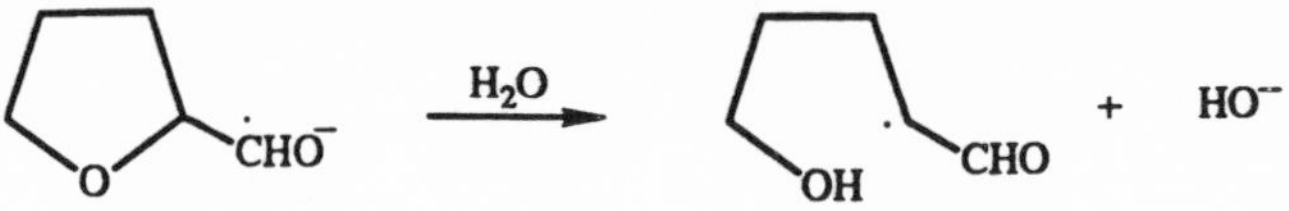

(19)

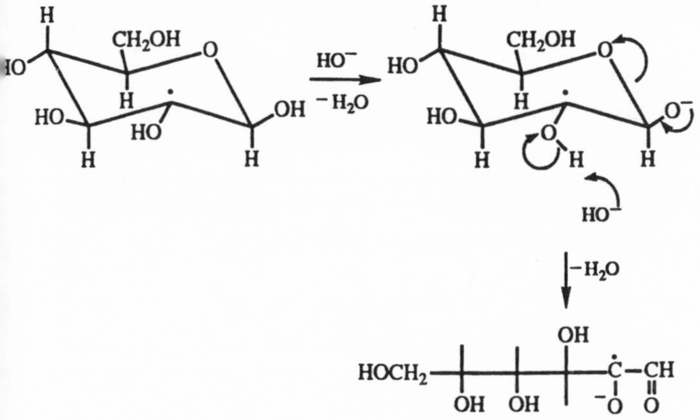

(20)

(13)

increased. In addition to the straightforward loss of OH^-, to give α-substituted radicals (see, e.g., Scheme 2), Steenken[15] has reported examples in which a β alkoxyl group is effectively lost [see, e.g., the ring-opening process shown in reaction (19)] and we have detected two types of *semidione* derived from sugars in related mechanisms involving both C-O and C–C fission. For example, the detection of radical **13** from β-D-glucose at pH circa 7 is attributed to the related fragmentation reaction (20) of the C_2-derived radical, and the detection of the three-carbon semidione (**15**) from β-D-glucose and ·OH at pH circa 9 is interpreted in terms of the anionic fragmentation of the ring-opened α-hydroxyalkyl radical **14** (see Scheme 3).

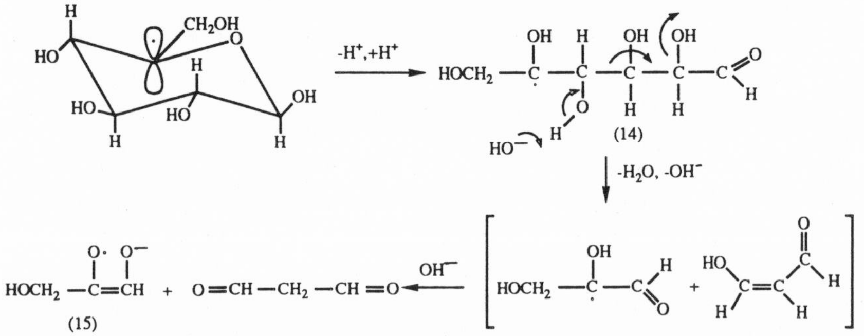

Scheme 3

4. RELATED HETEROLYTIC FRAGMENTATION REACTIONS: OTHER β LEAVING GROUPS

As might be expected on the basis of the results and mechanisms described, the occurrence of rearrangements from α hydroxyalkyl radicals is not restricted to the acid-catalyzed loss of OH^- or OR^- (and equivalent base-catalyzed mechanisms) but is also found to occur for other good anionic leaving groups.[6,20–23] Examples of this type of process [reactions (21) and (22)] include chloride, acetate, amine, and phosphate derivatives, examples of which will be discussed in turn, together with evidence for radical cations (e.g., **16**) for molecules for which α deprotonation is not possible.

$$·CH(OH)CH_2X \longrightarrow H^+ + ·CH_2CHO + X^- \tag{21}$$

$$·CH(OR)CH_2X \longrightarrow [CH(OR){=}CH_2]^{+·} + X^- \tag{22}$$
$$(16)$$

The α-hydroxyalkyl radical ·CH(OH)CH$_2$Cl from 2-chloroethanol is not detectable by ESR in the reaction of the latter with ·OH in aqueous solution,[20] but evidence for its formation comes from the detection of the rearranged radicals shown in Scheme 4a: loss of chloride is evidently faster than loss of OH⁻ from ·CH(OH)CH$_2$OH (which requires acid catalysis), and, as expected, ·CH$_2$CHO results by deprotonation of an intermediate radical cation. At higher pH, detection of ·CH(OH)CH$_2$OH and ·CH$_2$CH(OH)$_2$ provides evidence that rapid hydration of this intermediate can also occur, which also explains the observation of related behavior from 2-chloroethyl methyl ether (Scheme 4b): The corresponding radical from 2-methoxyethyl acetate [·CH(OMe)CH$_2$OAc] behaves similarly.[20]

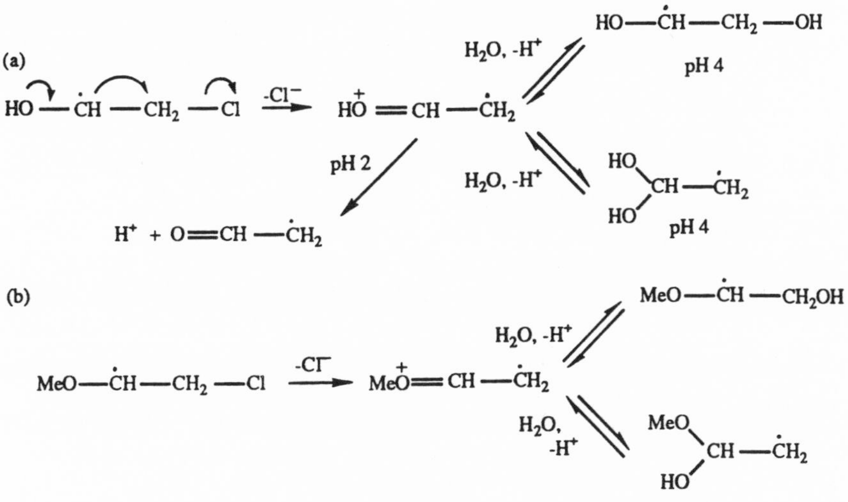

Scheme 4

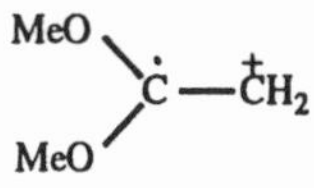

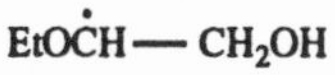

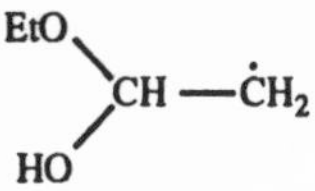

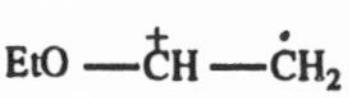

Other results that provide corroboration of the general mechanisms shown in Schemes 4a and 4b include the direct detection[23] of the ESR spectrum of the dialkoxy-substituted radical cation **17** (derived from loss of chloride from ·C(OMe)$_2$CH$_2$Cl under similar circumstances; the second alkoxy group stabilizes the radical cation and retards subsequent hydration: see refs. 23 and 24). Furthermore, when ·OH reacts with ethyl vinyl ether at pH circa 4 signals from **18** and **19**, with the former in excess, apparently reflect the kinetic control of addition to the alkene; however, as the pH is lowered, the predominance of **19** observed is believed to reflect an acid-catalyzed equilibration that brings about an OH shift, involving loss of the β OH and subsequent hydration of the radical cation **20** [isomer **19** being thermodynamically preferred]. Attack of **20** on the parent compound (CH_2=CHOEt), i.e., "spin trapping" of the radical cation, is believed to underlie the observation of **21** at very low pH.

A variety of more detailed and sophisticated ESR and pulse radiolysis/conductivity experiments have been performed on α-alkoxyalkyl radicals with β chlorine or acetoxy (leaving) groups.[22] For example, it is shown that the rate constants for hydrolysis of ·$CH(OMe)CH_2Cl$ and ·$CH(OMe)CH_2OAc$ (uncatalyzed) are $\geq 10^6$ s^{-1} and 2×10^3 s^{-1} at room temperature in aqueous solution, respectively (the reactions are much slower in acetone, so that ESR signals of the precursors can be obtained under steady state conditions). Other details described include the effect of alkyl substitution on the rate of loss of OAc [accelerated markedly by substitution at C_1 or C_2, as expected for a polar transition state characteristic of the proposed S_N1 type of reaction and as also described for phosphates (see below)]; the regioselectivity of attack by water on the intermediate radical cations (under both kinetic and thermodynamic control, at pHs 5–9 and ≤ 3, respectively); and the kinetics and selectivity of attack by other nucleophiles, including ¯OH and HPO_4^{2-} (see refs. 22 and 24).

Reaction of ethanolamine and related compounds reveals behavior broadly similar to those discussed earlier.[20] Thus although ethanolamine itself reacts with ·OH at pH circa 1 to give ESR signals from the radical ·$CH(OH)CH_2NH_3^+$ (the electrophilic hydroxyl radical would be expected to avoid the electron-withdrawing quaternary ammonium substituent), the N-trimethylated compound gives ·$CH(OH)CH_2NMe_3^+$ and the radical cation $^+NMe_3$, and $CH_2(OH)CH_2NEt_3^+$ gives solely $^+NEt_3$, consistent with a mechanism in which overall (apparent) homolytic fission [reaction (23)] is accelerated by the introduction of electron-donating substituents on nitrogen. On the other hand, a number of α-hydroxyalkyl groups with β ammonium or alkylammonium substituents have been observed[21] to undergo fragmentation to give α-ketoalkyl radicals (cf. the behavior of glycol and its analogs) at higher pH, a reaction that

$$\cdot CH(OH)CH_2NMe_3^+ \longrightarrow HOCH=CH_2 + \ ^{\cdot +}NMe_3 \tag{23}$$

$$\cdot CR^1(OH)CH_2\overset{+}{N}R_3^2 \ \underset{-H_2O}{\overset{HO^-}{\longrightarrow}} \ \cdot CR^1(O^-)CH_2\overset{+}{N}R_3^2 \ \longrightarrow \ O=CR^1CH_2^\cdot + NR_3^2 \tag{24}$$

is believed to involve the (ionized) ketyl radical [reaction (24); cf. Scheme 2]. [It has also been suggested[21] that the apparent anomaly between the homolytic and heterolytic fragmentations depicted in reactions (23) and (24) can be rationalized if reaction (23) occurs first to give an intermediate enol radical cation and amine, followed by single-electron transfer within a solvent cage so that the amine is subsequently oxidized.]

A more detailed pulse–radiolysis/conductivity and ESR study of the zwitterionic radicals $\cdot CH(OH)CH(CO_2^-)NH_3^+$ and $\cdot CMe(OH)CH(CO_2^-)NH_3^+$ from the amino acids serine and threonine reveals[25] some subtle differences in behavior, and evidence has been obtained for both a *ketyl*-type mechanism, for the former (signals from which are replaced by those from the secondary radical at pH > 4) and dissociation of the *neutral* α-hydroxyalkyl radical for the latter (at pH > 2) to give a radical zwitterion. These are illustrated in reactions (25) and (26), together with kinetic and acid/base information obtained from PR and ESR studies as a function of pH. Behrens and Koltzenburg have also discussed[25] the factors determining the type of mechanism for heterolytic fission in radicals of this type. They conclude that the +M effect of the α OH group is crucial, that methyl group substitution favors a changeover to radical cation (zwitterion) behavior [cf. reaction (26)] and that, in considering the potential leaving-group ability (nucleofugacity), it appears to be best measured via the pK of the corresponding acid (HCl > HOAc > NH_3).

$$\cdot CH(OH)CH(COOH)NH_3^+ \quad \xrightleftharpoons[\text{290 K}]{\substack{pK_a \ 2.2 \\ k \ 2.4 \times 10^6 \ s^{-1}}} \quad \cdot CH(OH)CH(COO^-)NH_3^+$$

$$NH_3 + \cdot CH(COO^-)CHO \qquad\qquad \cdot CH(O^-)CH(COO^-)NH_3^+ \quad \updownarrow \ pK_a \ 7.0 \tag{25}$$

$$\cdot CMe(OH)CH(COOH)NH_3^+ \quad \rightleftharpoons \quad \cdot CMe(OH)CH(COO^-)NH_3^+$$

$$\downarrow k \ 2.7 \times 10^4 \ s^{-1} \tag{26}$$

$$\cdot CH(COO^-)C(O)Me \quad \xleftarrow{-H^+} \quad \underset{Me}{\overset{HO}{>}}\!C^+ \!\!-\! \overset{H}{\underset{COO^-}{<}}\!C \ + \ NH_3$$

An example of a simpler amino-alcohol radical, in which both *ketyl* and *neutral* (radical cation) dissociation is revealed by steady state ESR studies, is shown in Scheme 5. The role of the β methyl groups in accelerating fragmentation of the neutral radical (cf. related diol-derived radicals discussed earlier)

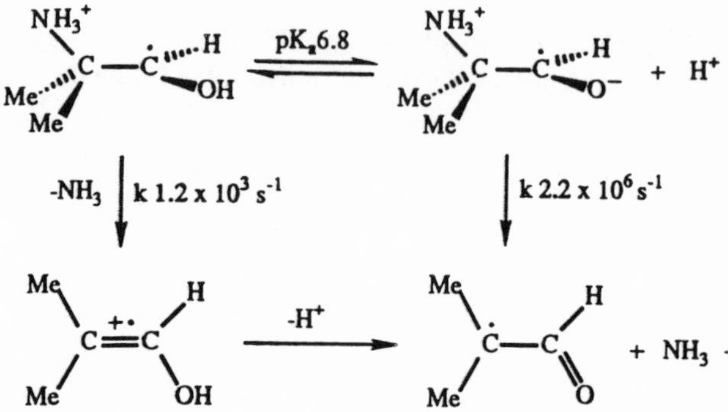

Scheme 5

is believed to reflect the (electronic) stabilization of the incipient radical cation and the (steric) effect that encourages eclipsing of the SOMO and β C–N bond.

Of special interest are a related series of rearrangements in which β phosphate groups are lost from, for example, glycerol phosphates [see e.g., reaction (27)],[26] sugar phosphates [see, e.g., reaction (28)],[27] and some 2-methoxyethyl phosphates [e.g., reaction (29)][28]; pulse–radiolysis experiments, as well as ESR spectra, have been reported.[26–28]

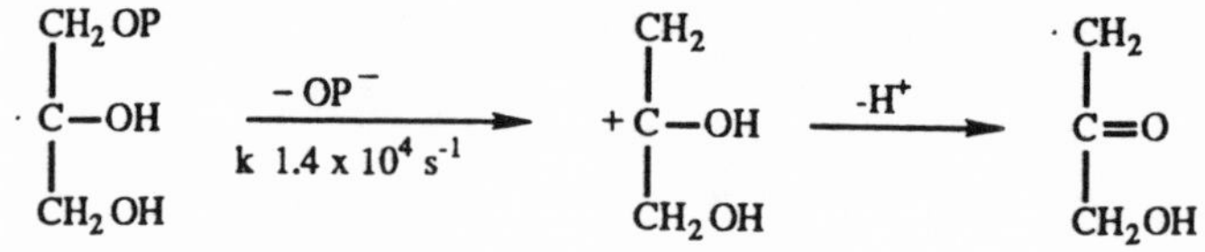

$OP = OP(O)(OH)O^-$

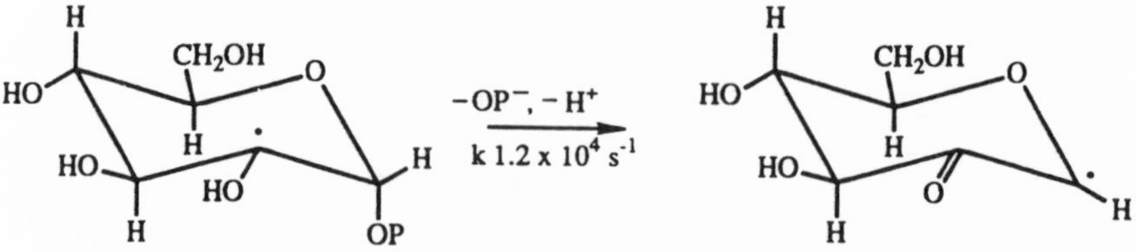

$$\cdot CH(OMe)CH_2OP \xrightarrow[H_2O, \; -H^+]{-OP^-} \cdot CH(OMe)CH_2OH + HOCH(OMe)\dot{C}H_2 \quad (29)$$

The rate constants for analogs of reaction (29) involving differently ionized and alkylated β substituents (see Table 2) reveal that $H_2PO_4^-$ is a much better leaving group than HPO_4^{2-} and PO_4^{3-} (as would be predicted from the pK_a

Table 2.   Rate Constants for Phosphate Elimination from Some
Oxygen-Conjugated β-Phosphate-Substituted Radicals

Parent Radical	k/s^{-1}	Temperature ($°C$)	Ref.
MeOĊHCH$_2$OPO$_3^{2-}$	0.1–1	0	28
MeOĊHCH$_2$OPO$_3$H$^-$	circa 10^3	0	28
MeOĊHCH$_2$OPO$_3$H$_2$	circa 3×10^6	0	28
MeOĊHCH$_2$OP(O)$_2$(OCH$_2$CH$_2$OMe)$^-$	10^3–10^4	0	28
MeOĊHCH$_2$OP(O)(OH)(OCH$_2$CH$_2$OMe)	circa 2×10^7	0	28
MeOĊHCH$_2$OP(O)(OCH$_2$CH$_2$OMe)$_2$	$>4 \times 10^7$	0	28
·CH(OH)(CH$_2$OH)CH$_2$OPO$_3$H$^-$	1.4×10^4	20	27
α-D-Glucose-1-phosphate-2-yl	1.2×10^4	20	27
Fructose-1,6-diphosphate-5-yl	10^2–10^3	20	27
Ribose-5-phosphate-4-yl	$>10^3$	20	27

values of the corresponding acids, as described previously) and show that alkylation of the phosphate serves to accelerate the reaction. It is also found that a correlation exists between the rate constant for elimination and the values of the splitting constants (a_P and $a_{β-H}$) in the ESR spectra of the appropriate radicals: this is believed to reflect a relation (referred to earlier) between ease of loss of a β group and its –I effect, which in turn encourages (electronic) interaction with the SOMO, hence causing favored adoption of an eclipsing geometry (revealed by the splitting: see also ref. 14).[28]

As noted above, in the reaction of ·OH with a range of sugar phosphates elimination of phosphate follows attack to produce a radical center adjacent to the carbon bound to phosphate—an initial mode of attack preferred for fructose 1,6-diphosphate, fructose 6-phosphate, β-D-fructofuranose itself, and, to a lesser extent, ribose-5-phosphate (a selectivity, absent in pyranose sugars, that reflects stereoelectronic features discussed previously). These ESR observations and the results discussed above (e.g., for alkylated phosphates) lend considerable support to the proposal[5,6,29] that an important mechanism for radical-induced strand breakage in DNA involves analogous hydrogen abstraction (e.g., at the carbon attached to the oxygen in the furanose ring, either directly by ·OH or via nucleo-base-derived radicals) followed by rapid cleavage of the β-(C)-phosphate bond (C$_5'$ or C$_3'$).

Although most of the information concerning the mechanism of such processes (including the evidence for predominant initial attack of ·OH at the bases in DNA) derives from pulse–radiolysis and product studies, ESR spectroscopy does allow very direct evidence to be obtained for related processes in sugar-derived radicals[30]: examples are provided in reactions (30)–(32). Thus, evidence has been obtained of reaction of ·OH at C$_4$ and C$_2'$ of adenosine: for

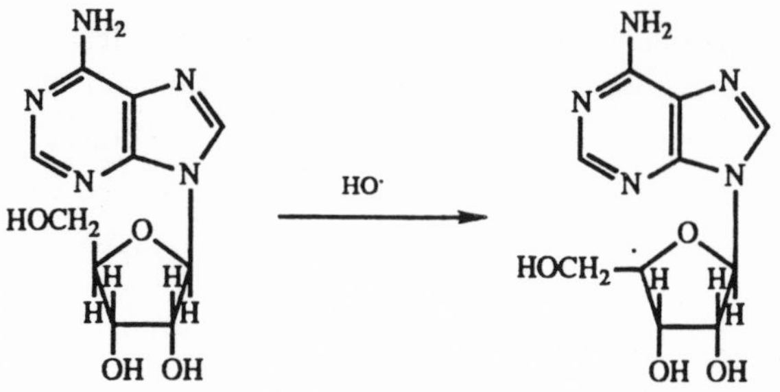

(30)

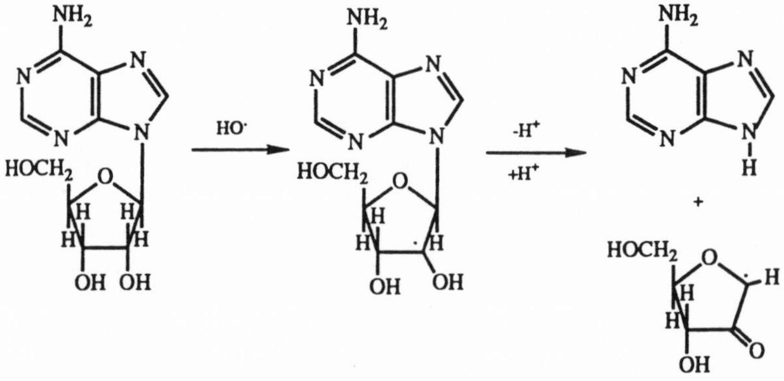

(22)

(31)

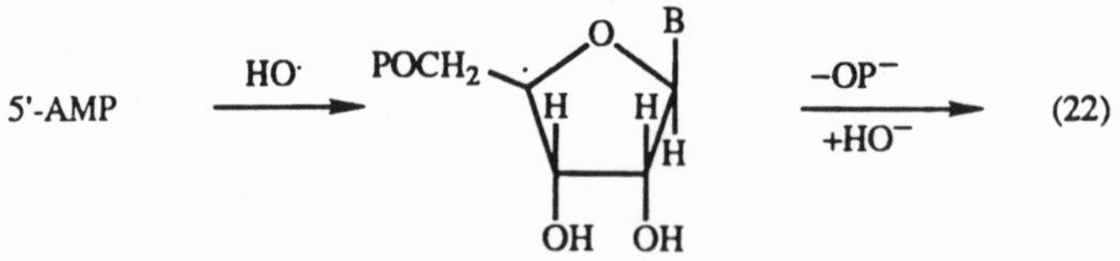

(23)

(22)   (32)

the latter route, the detection of **23** rather than the precursor indicates that loss of the β leaving group (here the free base adenine) has occurred, presumably via a mechanism related to those discussed earlier. The observation of the signal from **22** from adenosine-5-monophosphate (5'-AMP) indicates that loss of phosphate ($k > 10^3$ s^{-1}) has followed attack at C_4', an observation that supports arguments for strand breakage in DNA via attack at C_4'. Evidence has also been presented for rapid cleavage of the β C_3'-phosphate group in C_2'-derived radicals from polyuridylic acid.[29c,d]

5. EXAMPLES OF 1,2 SHIFTS IN RADICALS IN AQUEOUS SOLUTION

The interconversion of the two hydroxyl adducts from ethyl vinyl ether ·CH(OEt)CH$_2$OH and ·CH$_2$CH(OH)OEt—albeit in acid solution only, and probably via dissociation/recombination—constitutes an example of a 1,2 shift in a radical. Though a small number of 1,2 shifts in radicals have been described[4]—e.g., in radicals of the type ·CH$_2$CH$_2$X (X = Cl), including those believed to proceed via addition–elimination sequences (e.g., for X = SR)— they are uncommon under the conditions we have used for most of the studies described here. These particular questions were of interest to us in our search for more examples: is a truly *intra*molecular 1,2-OH shift observable for radicals in solution? Can *unactivated* radicals of the type ·CR$_2^1$CR$_2^2$OH be encouraged to undergo rearrangement in acid conditions? Is the rearrangement of ·CH$_2$CMe$_2$OAc into ·CMe$_2$CH$_2$OAc (see ref. 31) related to those discussed above?

It has been clearly established that reaction of ·OH with *tert*-butyl alcohol at pH circa 1 gives almost exclusively the radical ·CH$_2$CMe$_2$OH (**24**), detectable by ESR spectroscopy in Ti^{III}/H$_2$O$_2$ and Fe^{II}/H$_2$O$_2$ systems. However, in similar experiments in which the pH is reduced to circa 0, the spectrum from a second species becomes detectable; it is assigned to the isomeric species ·CMe$_2$CH$_2$OH (**25**).[32] It has been proposed[32] that this acid-catalyzed reaction proceeds via the formation of an intermediate radical cation [see reaction (33)], though *ab initio* calculations[33] on the possibility of a 1,2 shift in ·CH$_2$CH$_2$OH and its protonated counterpart indicate that protonation would be expected to lower significantly the energy of the transition state for an intramolecular 1,2 shift.

$$\cdot CH_2CMe_2OH \quad \xrightarrow[-H_2O]{H^+} \quad [CH_2=CMe_2]^{+\cdot} \quad \xrightarrow{H_2O, \, -H^+} \quad HOCH_2\overset{\cdot}{C}Me_2 \qquad (33)$$

$$\text{(24)} \qquad\qquad\qquad\qquad\qquad\qquad\qquad\qquad\qquad \text{(25)}$$

Other aspects of this type of reaction have been explored in some detail in an attempt to establish its nature and the factors that govern its occurrence. For example, the observation that 3-ethylpentan-3-ol (Et$_3$COH) reacts with ·OH to give ·CHMeC(OH)Et$_2$ (**26**) and ·CH$_2$CH$_2$C(OH)Et$_2$ (**27**) at pH > 2.5 (Scheme 6) is as would be expected; however, as the pH is lowered the former, but not the latter, is removed, to be replaced by ESR signals from its isomer ·CEt$_2$CH(OH)Me (**28**), which in turn is replaced by signals from ·CHMe–CEt=CHMe (**29**) at even lower pH (<0). These observations can be explained if protonation at oxygen of the β-hydroxyalkyl radical (**26**) leads to loss of water and formation of a radical cation (**30**) (see Scheme 6) that can undergo

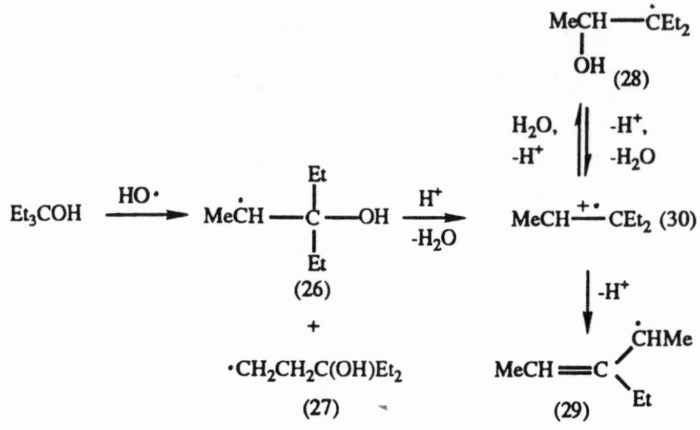

$$MeCH\!-\!\overset{\bullet}{C}Et_2$$
$$\overset{|}{O}H \quad (28)$$

$$H_2O, \quad \Big|\Big| \quad -H^+,$$
$$-H^+ \quad \Big|\Big| \quad -H_2O$$

$$Et_3COH \xrightarrow{HO\cdot} Me\overset{\bullet}{C}H\!-\!\underset{\underset{Et}{|}}{\overset{\overset{Et}{|}}{C}}\!-\!OH \xrightarrow[-H_2O]{H^+} MeCH\overset{+\bullet}{-\!\!-}CEt_2 \;(30)$$

$$(26)$$

$$+ \qquad\qquad\qquad\qquad \Big| -H^+$$

$$\cdot CH_2CH_2C(OH)Et_2 \qquad MeCH\!=\!C\overset{\overset{\bullet}{C}HMe}{\underset{Et}{\diagdown}}$$

$$(27) \qquad\qquad\qquad\qquad (29)$$

Scheme 6

reversible hydration [to lead to the more stable radical (28) or, ultimately, react irreversibly via deprotonation to give the stabilized alkyl radical (29)].

A variety of other β-hydroxysubstituted radicals have been explored and it has been shown that the rate of this rearrangement (as measured by the pH at which the ESR spectra of new radicals begin to replace those of the first-formed radicals in steady state experiments) appears, as might be expected, to be related to the ionization potential of the alkene whose radical cation is proposed as an intermediate. Thus rearrangement of $\cdot CH_2CMe_2OH$ occurs at pH circa –0.5 whereas $\cdot CMe(OH)CH_2OH$ undergoes acid-catalyzed rearrangement at pH circa 4 (ionization potentials of $CH_2=CMe_2$ and $MeC(OH)=CH_2$ are circa 9.2 and 8.2 eV, respectively).

More direct evidence that radical cations are involved in these processes is provided by pulse–radiolysis studies (in conjunction with ESR investigations) of the acid-catalyzed elimination of OH^- from $\cdot CMe_2CMe_2OH$ (31) [generated by addition of $\cdot OH$ to tetramethylethylene (32) and by hydrogen abstraction from 2,3-dimethylbutane-2-ol (see Scheme 7)].[34] With 32, absorption spectra of two intermediates could be detected shortly after the pulse (and formation of $\cdot OH$) in acid conditions. An absorption with λ_{max} 240 nm (Figure 8) detected when the pH is lowered below 2.5 is assigned to the radical cation (33); at lower pH (<0.3) the second intermediate detected (with λ_{max} 290 nm) is assigned to the allyl radical $\cdot CH_2CMe=CMe_2$ formed by deprotonation. Kinetic analysis of the results yields values for the deprotonation of the radical cation $(3.9 \times 10^5$ $s^{-1})$ and the equilibrium constant for interconversion of 31 and 33 (1.2). For the analogous reactions of Et_3COH, the rate constant for deprotonation of $[Et_2C=CHMe]^{+\cdot}$ is calculated to be $2.5 \times 10^5\,s^{-1}$, with $K = 2.5$ for the corresponding hydration equilibrium involving $\cdot CHMeCEt_2OH$.

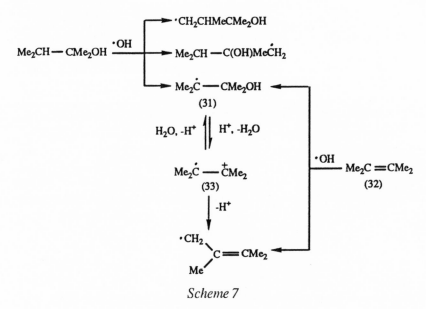

Scheme 7

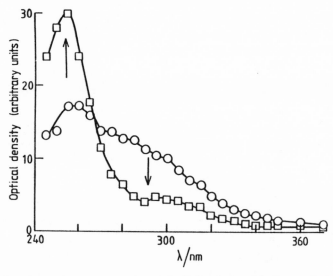

Figure 8. Absorption spectra of $[Me_2C=CMe_2]^+$ and $\cdot CH_2CMe=CMe_2$ respectively recorded 0.5 μs (O) and 10 μs (□) after pulse radiolysis of an aqueous solution of $Me_2C=CMe_2$ at pH circa 0 (see Scheme 7).

There has been continued interest in the detailed mechanism whereby β-acetoxy-substituted alkyl radicals undergo a formal 1,2-shift in aqueous solution at room temperature [see, e.g., reaction (34), for the radical derived by hydrogen-atom abstraction from *tert*-butyl acetate, for which a rate constant of circa 10^3 s^{-1} may be estimated at room temperature].[31]

$$\text{·CH}_2\text{CMe}_2\text{OCMe} \longrightarrow \text{MeCOCH}_2\text{·CMe}_2 \qquad (34)$$

In an elegant series of studies (see, e.g., refs. 4, 31, and 35 and references therein) it has been clearly demonstrated that reaction does not proceed via the cyclic dioxolan-2-yl free radical (e.g., **34**), since such species do not undergo ring-opening under these conditions (indeed, spectra of the intact cyclic radicals can be detected by ESR when generated from cyclic precursors under similar conditions). This reaction, at least for simple acyclic analogs, is believed to be concerted and to proceed via a five-membered cyclic transition state [see, e.g., **35**] to give products in which the carbonyl and ether oxygens of the starting material are interchanged. However, more recent kinetic and ^{18}O-labeling studies have established that, depending in particular on the structure of the β-(acyloxy)alkyl radical involved, a different mechanism can be followed[35,36]: thus, retention of the ^{18}O carbonyl label in this function in the 1,2-rearrangement of **36** is interpreted in terms of a more rapid rearrangement either through a three-membered transition state or, more likely, a tight anion/radical cation ion pair (see **37**).

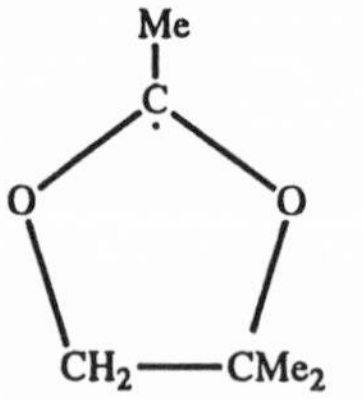

(34) 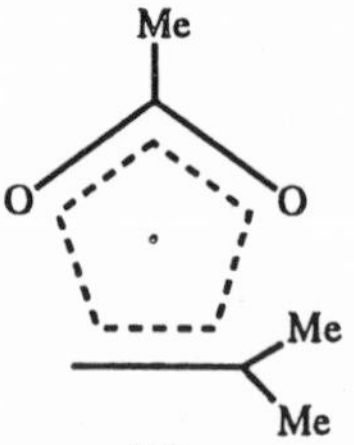   (35)

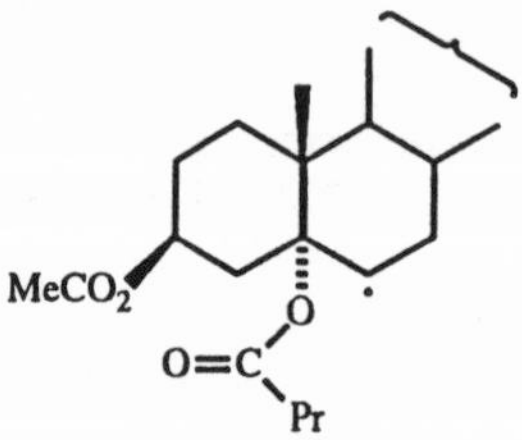

(36) 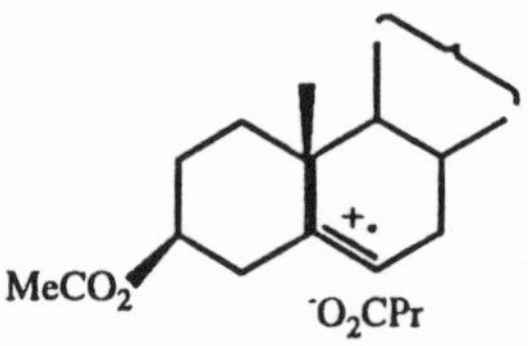   (37)

The rearrangement of $\cdot CH_2CMe_2OC(O)CF_3$ into $CF_3C(O)OCH_2\dot{C}Me_2$ is considerably faster than reaction (34) under comparable conditions ($k_{75°C} = 7 \times 10^4$ s^{-1} in hydrocarbon solvents).[37] This, and the acceleration of the rearrangement brought about by the incorporation of a cyano group in the acetate function [for $\cdot CH_2CMe_2OC(O)CH_2CN$], provides further evidence for a charge-separated transition state.

6. ALKENE RADICAL CATIONS

The evidence presented above for the isomerization of some β-hydroxyalkyl radicals (such as **24** into **25**) in strongly acidic solution[32] and the correlation between the ease of loss of OH^- and the stability of the proposed radical cation intermediates (as measured by the ionization potentials of the corresponding alkenes) suggested to us that similar transient radical cations might be observed on oxidation of the parent alkenes (provided suitable electron-donating substituents were present) with either $HO\cdot$ in acidic solution, or other powerful one-electron oxidants such as SO_4^{-} and Cl_2^{-}. The former species can be generated either by reaction of Ti^{III} with $S_2O_8^{2-}$ [reaction (35)][38] or direct photolysis of $S_2O_8^{2-}$ [reaction (36)],[39] while the latter species is formed on reaction of either $HO\cdot$ or SO_4^{-} with excess Cl^- [reactions (37) and (38)].[40]

$$Ti^{III} + S_2O_8^{2-} \longrightarrow Ti^{IV} + SO_4^{-}\cdot + SO_4^{2-} \tag{35}$$

$$S_2O_8^{2-} \xrightarrow{h\upsilon} 2SO_4^{-}\cdot \tag{36}$$

$$HO\cdot + Cl^- \longrightarrow HOCl^{-} \underset{}{\overset{H^+}{\rightleftharpoons}} H_2O + Cl\cdot \tag{37}$$

$$SO_4^{-}\cdot + Cl^- \longrightarrow SO_4^{2-} + Cl\cdot \xrightarrow{Cl^-} Cl_2^{-}\cdot \tag{38}$$

The results obtained in such studies establish that while addition of these radicals to the alkenes is the major pathway with unsubstituted or mono-alkylated alkenes [see, for example, reactions (39)–(41)], with highly substituted alkenes and certain dienes overall one-electron oxidation (and subsequent hydration) is effected.[41] Thus, reaction of Cl_2^{-} with 2-methylpropene gives rise to ESR signals from $\cdot CMe_2CH_2OH$ rather than $\cdot CMe_2CH_2Cl$, and reaction of either SO_4^{-} or Cl_2^{-} with $Me_2C=CH–CO_2H$ gives rise to signals from both $\cdot CMe_2CH(OH)CO_2H$ and $\cdot CH(CO_2H)CMe_2OH$.[41]

$$SO_4^{-}\cdot + CH_2=CH_2 \longrightarrow {}^-OSO_3CH_2CH_2\cdot \tag{39}$$

$$SO_4^{-}\cdot + CH_2=CHR \longrightarrow {}^-OSO_3CH_2\dot{C}HR \tag{40}$$

$$Cl_2^{-}\cdot + CH_2=CHR \longrightarrow Cl^- + ClCH_2\dot{C}HR \tag{41}$$

6.1. Hydration Processes

The formation of species such as **38** or **39** could arise via three different routes: direct electron transfer from the alkene to SO_4^- and Cl_2^- and subsequent hydration, for example, by reaction (42); reaction via initial *adduct* formation (as observed for the less substituted alkenes) followed by S_N1 hydrolysis [i.e., via hydration of the radical cation; reaction (43)]; or reaction via adduct formation and S_N2 hydrolysis [a mechanism that would involve a polar transi-

$$Me_2C=CHCOOH \xrightarrow{SO_4^-/Cl_2^-} [Me_2C=CHCOOH]^{+\cdot} \xrightarrow[-H^+]{H_2O} \cdot CMe_2CH(OH)COOH \quad (38)$$

$$+$$

$$Me_2C(OH)\dot{C}HCOOH$$
$$(39)$$

(42)

$$Me_2C=CHCOOH \xrightarrow{SO_4^-} \left[\begin{array}{c} \cdot CMe_2CH(OSO_3^-)COOH \\ Me_2C(OSO_3^-)\dot{C}HCOOH \end{array} \right] \xrightarrow{-SO_4^{2-}} [Me_2C=CHCOOH]^{+\cdot} \qquad (43)$$

$$\downarrow$$

$$(38) + (39)$$

$$Me_2C=CHCOOH \xrightarrow{SO_4^-} \cdot CMe_2CH(OSO_3^-)COOH \quad + \quad Me_2C(OSO_3^-)\dot{C}HCOOH \qquad (44)$$

$$\quad\quad\quad\quad\quad\quad\quad\quad \downarrow {\scriptstyle H_2O \atop -H^+,-SO_4^{2-}} \quad\quad\quad\quad\quad\quad \downarrow {\scriptstyle H_2O \atop -H^+,-SO_4^{2-}}$$

$$\quad\quad\quad\quad\quad\quad\quad (38) \quad\quad\quad\quad\quad\quad\quad\quad (39)$$

tion state and avoid the formation of a discrete radical cation intermediate; reaction (44)]. For each of these mechanisms, the incorporation of electron-donating groups would be expected to facilitate the reaction: the lowering of the ionization potential would clearly assist *direct* electron transfer but would also assist subsequent heterolytic (S_N1) fragmentation of an adduct by stabilization of the resulting radical cation as well as the polar transition state of S_N2 hydrolysis. Heterolytic cleavage might also be aided by steric effects on the conformation of the intermediate.

Indirect evidence against the occurrence of both the S_N2 hydrolysis and direct electron transfer pathways was obtained from a study of the reaction of SO_4^- with 3,3- and 2,3-dimethylacrylic acid.[38,41] These compounds, which would be expected to have very similar ionization potentials, behave in a very different manner, with the former giving rise to hydroxylated species and the latter only the SO_4^- adduct $\cdot CMe(CO_2H)CH(OSO_3^-)Me$. This difference can, however, be explained in terms of the steric acceleration of heterolytic S_N1 fragmentation by the gem dimethyl groups in the first-

$$\underset{\text{(40)}}{\overset{\displaystyle ^-O_3SO}{\underset{\underset{Me}{\overset{|}{\underset{\displaystyle Me}{}}}}{\overset{\displaystyle |}{\underset{\displaystyle \;}{C}}}-\overset{\displaystyle \overset{\frown}{C}\overset{H}{\underset{\displaystyle CO_2H}{}}}{\;}}$$

formed *adduct* $\cdot CH(CO_2H)CMe_2(OSO_3^-)$, in which the preferred conformation (40) would be expected to possess an eclipsing geometry of the β C–O bond and the unpaired electron.

Investigations of the reaction of radiolytically generated HO· radicals with alkyl bromides, chlorides, methanesulfonates, dialkylsulfates and trialkylphosphates lend credence to the proposed addition/hydration pathway. Thus, reaction of HO· with, for example, 1-chloro-2-methylpropane gives rise to the rapid formation of acid, which can be detected by time-resolved conductivity measurements[42]: the observed increase in conductivity has been assigned to the hydrolysis of the initially formed radical $\cdot CMe_2CH_2Cl$ [reaction (45)]. In analogous reactions the rate constant was found to be highly dependent on the

$$^\cdot CMe_2CH_2Cl \; + \; H_2O \; \longrightarrow \; ^\cdot CMe_2CH_2OH \; + \; Cl^- + \; H^+ \qquad (45)$$

structure of the initial radical, the values increasing with increasing alkyl substitution. These observations, together with the finding that the acceleration in rate is greatest when the alkyl substituent is introduced in the α position, suggests an S_N1 type of hydrolysis involving the formation of an alkene radical cation intermediate. Attempts to scavenge these species with added HPO_4^{2-} ions were not, however, successful, suggesting that the half-lives of the radical cations (if they are fully developed as discrete species) are shorter than 10^{-10} s in aqueous solution at room temperature.[42]

Recent product studies on the hydroxylation of substituted cyclohexenes induced by $SO_4^{\overline{\cdot}}$ provide further evidence for the existence of discrete radical cations in these reactions.[43] The radical species **41** and **42** produced on reaction of HO· with the parent sulfate (R = *tert*-butyl) have been shown to yield, in the presence of H donors, the corresponding substituted cyclohexanols as products (**43–46**). Quantification of the yields of the four possible isomeric products shows that products with the axial OH groups dominate, and that a high degree of inversion occurs. Thus radical **41**, with a cis configuration, gives 56% of the alcohols with a trans configuration and **42**, with a trans configuration, gives 59% of the alcohols with a cis configuration. These observations, together with the fact that radicals **41** and **42** do not give the same isomer distribution (as would have been expected if a completely free radical cation were to be generated) are in agreement with the proposal that the reaction involves a partially solvent-separated radical cation/sulfate ion pair (**47**) as an inter-

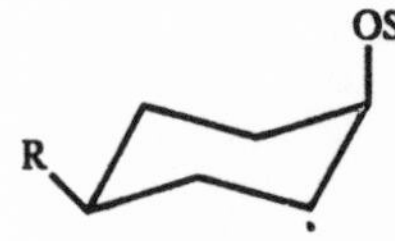

(41)    (42)

(43) 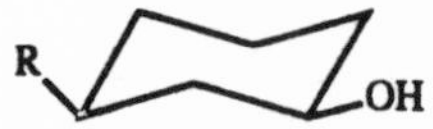   (44) 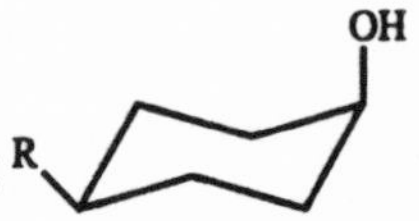   (45)    (46)

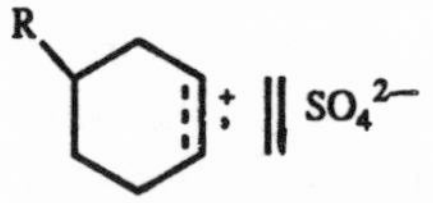

(47)

mediate, attack of water on the radical cation occurring before the sulfate group has completely departed.[43]

The hydration of such radical cation/sulfate ion pairs has been shown to be remarkably stereoselective when compared with direct radical hydroxylation brought about by HO· radicals. Thus the ratios of the products **43** : **44**, where R = Me$_2$COH, and **45** : **46** are both circa 11 : 1 when obtained from SO$_4^-$ reaction with the alkene whereas direct HO· attack gives circa 1 : 1.[43] Such selectivity in hydration presumably reflects preferential formation of the hydroxylated species with an eclipsing geometry of the β C–O bond and the unpaired electron. Hydration of acyclic radical cations (or radical cation/anion pairs) appears on the other hand to be less selective than HO· attack. Thus reaction of HO· with 3,3-dimethylacrylic acid at pH 2 gives the radicals **38** and **39** in the ratio circa 9 : 1 whereas oxidation by SO$_4^-$ or Cl$_2^-$ gives these same adducts but in the ratio circa 3 : 1 (see Scheme 8).[41] This latter ratio presumably reflects kinetic control of hydration whereas the former reflects kinetic control of hydroxyl radical addition. In the reaction of HO· with this

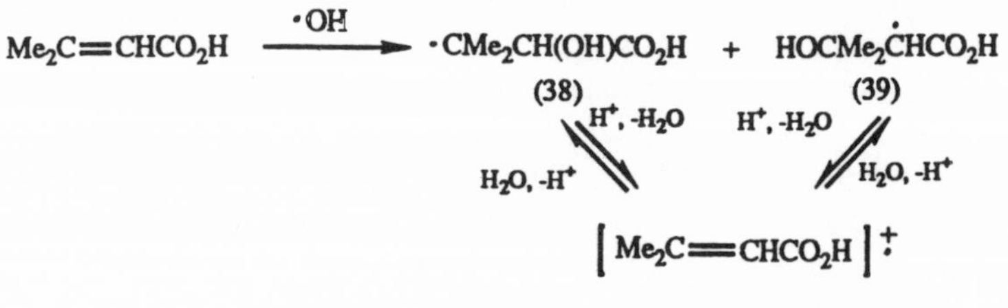

Scheme 8

substrate, the ratio of the adduct species decreases as the pH is lowered until at pH circa 0 it is similar to that observed with SO_4^- and Cl_2^-. This changeover presumably reflects acid-catalyzed loss of HO^- and interconversion of the isomers via the radical cation, though whether complete equilibration has been achieved (and hence thermodynamic control of the ratio of the product radicals) is not clear; even if this is not so, hydration appears to be *less* selective than HO· attack.

6.2. Dimerization Reactions

As might be expected from the findings reported on the lifetime of some simple alkene radical cations,[43] there are few examples of reactions that can compete with reaction with the solvent (H_2O). These competing processes are all unimolecular in nature with a single exception. This exception occurs with the conjugated diene muconic acid ($HO_2CCH=CHCH=CHCO_2H$) whose reaction with HO· at very low pH (circa 0) gives rise to a "dimeric" species (**48**) (in which R has a single proton in the position formally γ to the site of the unpaired

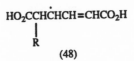

(48)

electron); this species is not observed at higher pH values [where the allyl species $HO_2C–CH(OH)–\dot{C}H–CH=CH–CO_2H$ predominates] even at higher concentrations of muconic acid, suggesting that it does not arise via addition of the allyl radical to the parent molecule.[41] This behavior resembles the reactions of other electron-rich monomers (e.g., $CH_2=CHOR$ and R_2S), which give dimeric species at very low pH values [e.g., $HO(RO)CHCH_2CH_2\dot{C}HOR$ and $R_2SSR_2^+$, respectively (see Section 9)] via reaction of a first-formed radical cation with the parent molecule; it seems likely that **48** is formed in a similar manner.[41]

6.3. Deprotonation Reactions

Loss of a β proton, a commonly observed process with certain other radical cations from electron-rich substrates has been suggested to occur with cyclopentene on oxidation with Cl_2^- to account for the observation of the radical **49** [reaction (46)]. In contrast, oxidation of this same substrate with SO_4^- gave only weak signals that are believed to be due to the adduct species **50** [reaction (47)].[41]

The increased rate of proton loss (relative to hydration) with this substrate compared to the acyclic species may arise from the adoption by the transient intermediate of a conformation that favors such fragmentation (i.e., with the β C–H bond pseudo-axial).

(46)

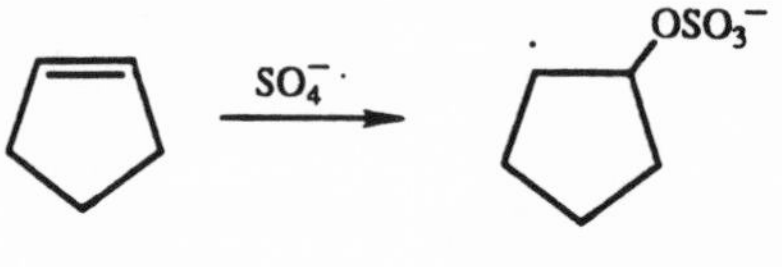

(49)

(47)

(50)

6.4. Intramolecular Processes

Attempts to scavenge these radical cations with species such as HPO_4^{2-} have proved unsuccessful[42,43] though such processes have been shown to occur with longer-lived intermediates such as α dialkoxyalkene radical cations.[24] In contrast, the observation of cyclized species such as **51** on oxidation of pent-4-en-1-ol (Scheme 9) and **52** on oxidation of hex-4-en-1-ol with Cl_2^-, HO· (at low pH), and SO_4^- (with the latter substrate only; reaction with the former gave only adduct species, a finding that is in agreement with results obtained with, for example, propene[41]) is interpreted as being due to internal nucleophilic attack of the hydroxyl group on the transient radical cation (see Scheme 9).[41] The regioselectivity of this process (which yields mainly **51** from pent-4-en-1-ol rather than **53**) resembles that of attack of the corresponding alkoxyl radical on the double bond[44]; this is not unexpected, as in both cases it would reflect the achievement of good overlap between the (incipient) unpaired electron on the

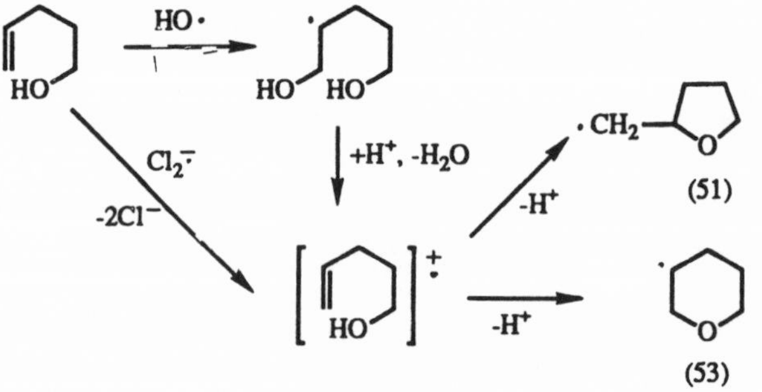

Scheme 9

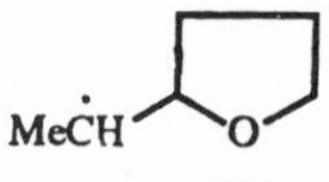

(52)

radical center and the developing β C–O bond. Production of alkoxyl radicals either directly by O–H abstraction, or via internal electron transfer from the oxygen to the radical cation (and subsequent rapid deprotonation) has been ruled out on the basis of the known chemistry of the attacking species and the ionization potentials of the alkene and alcohol functions.[45,46]

Comparison of these results with those obtained with the corresponding simple alkenes[41] indicates that *adduct* species might have been expected; furthermore, in the acid-catalyzed hydroxyl radical reactions, ring closure occurs at a higher pH than would have been predicted. These observations[41] may indicate the participation of a neighboring group effect in the departure of the leaving group [reaction (48); X = OH_2^+, Cl, OSO_3^-] and is consistent with the evidence that these incipient radical cations may never become discrete species.[43]

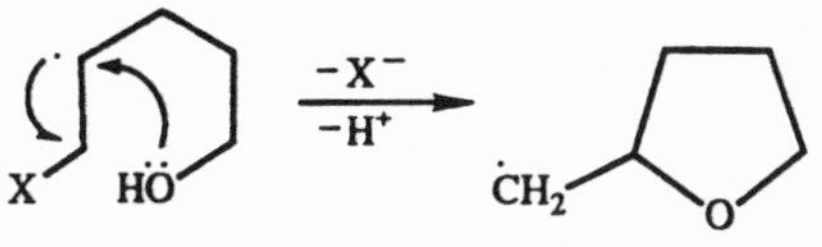

$$\text{(51)} \qquad (48)$$

Further studies have shown that similar internal nucleophilic attack can occur with carboxylic acids and their anions, and that the regioselectivity of these ring closures is primarily 1,5 (exo) [see, for example, reaction (49)][47]; such a geometry is as expected, as these reactions can be viewed as the reverse of the heterolytic cleavage of β OH or OR bonds mentioned earlier, for which the requirement for an eclipsing geometry has been established.

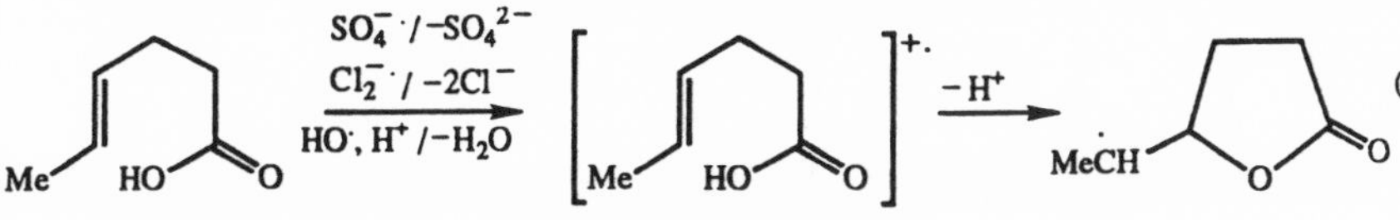

$$\text{(49)}$$

7. AROMATIC RADICAL CATIONS

As might be expected on the basis of their ionization potentials[46] a large number of aromatic compounds are also susceptible to one-electron oxidation with reagents such as SO_4^-, Cl_2^-, and HO· (at low pH). Continuous-flow ESR experiments show that reaction with ·OH leads initially to the formation of adducts (hydroxycyclohexadienyl radicals).[48–50] However, at low pH, radicals produced by side chain oxidation are observed; these observations can be rationalized in terms of the acid-catalyzed formation of arene radical cations, which, depending on structure, can undergo a number of different processes that are in some cases formally similar to those outlined above for alkenes. Similar side chain oxidation processes can be observed with SO_4^-

and Cl_2^{-}, though no evidence is yet available as to whether such oxidation processes proceed via *direct* one-electron abstraction or via adduct radicals that undergo extremely rapid heterolytic cleavage (cf. the behavior of alkenes); in many of these cases rapid hydration of the radical cation can also be observed. The major pathways that compete with hydration are outlined in the following sections.

7.1. Deprotonation to Form Benzylic Radicals

This is the major type of process observed for aromatic compounds with short alkyl chain substituents and for compounds such as benzyl alcohol and 1-phenylethanol.[50,51] The rate of deprotonation of these compounds appears to depend markedly on the substituents present and determines whether this process competes successfully with hydration. Thus oxidation of toluene with SO_4^{-} gives mainly hydroxycyclohexadienyl species (i.e., via hydration) at pH values greater than $2^{48,49}$; it is only at lower pH values, where rapid equilibration of these species with the radical cation occurs, that the benzyl radical is observed. In contrast, deprotonation is observed throughout the pH range 0.5–9.5 for benzyl alcohol, suggesting that in this case the rate of deprotonation is faster than that of hydration (see Scheme 10)[50]; a value of circa 10^9 s^{-1} has been estimated[50] for the rate constant for the former process on the basis of the ratio of the rate constants for deprotonation (circa 10^3) and a previously reported value for the deprotonation of the toluene radical cation of 10^6 s^{-1}.[49] This rate enhancement has been interpreted in terms of the more favorable formation of a stabilized oxygen-conjugated radical. The rate of loss of water from the initial hydroxyl adducts has also been estimated using a steady state analysis (a value of 8×10^5 dm^3 mol^{-1} s^{-1} is obtained for the loss of water from the hydroxycyclohexadienyl radical(s) shown in Scheme 10).

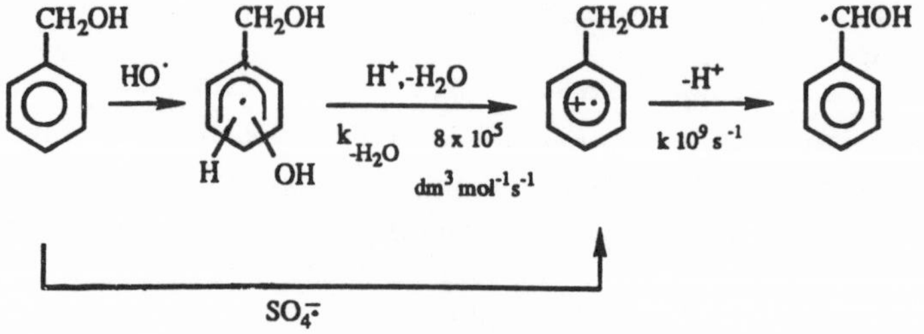

Scheme 10

7.2. C_α–C_β Fragmentation

With certain substrates this type of fragmentation competes effectively with deprotonation; as expected this is generally observed when one or both of the fragments is stabilized.[50,51] Thus oxidation of $PhCH_2CH_2OH$ with HO· gives, at pH values greater than 2.2, ESR signals from the hydroxycyclohexadienyl species, while at pH values below this signals assignable to the benzyl radical are observed[50]: this radical is also observed on reaction of SO_4^{-} with this alcohol at pH 1. This behavior is consistent with the proposal[52] that the first-formed radical cation can undergo both hydration and fragmentation (see Scheme 11) to give the benzyl radical and, presumably, $^+CH_2OH$ (favored by the stability of the latter carbonium ion).

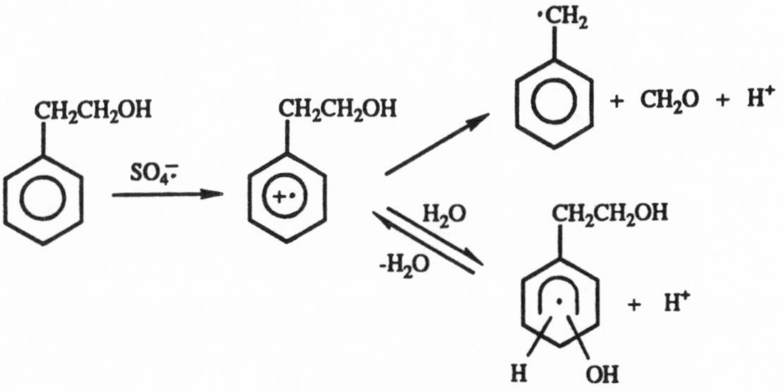

Scheme 11

In contrast, oxidation of $PhCMe_2OH$ gives rise to signals assigned to the methyl radical at pH values of <3.5,[50] which suggests that in this case the production of the highly stabilized carbonium ion $^+CPhMeOH$ [and hence $PhC(O)Me^{51}$] is the crucial factor [reaction (50)].

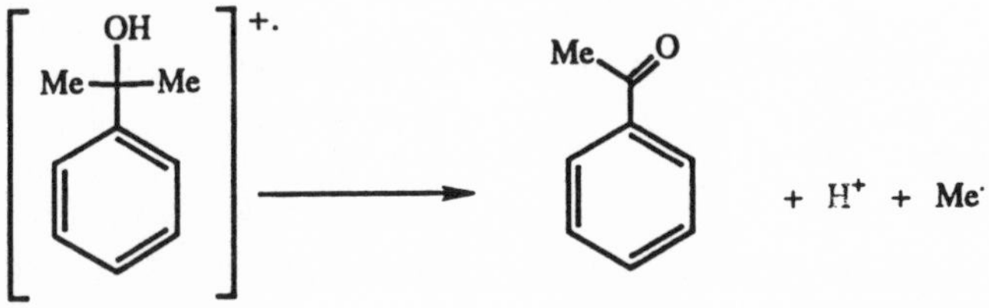

(50)

The pH values at which fragmentations are seen in reactions with HO· (2.5 for the formation of PhĊHMe from PhĊHMeCH_2OH, 3.2 for the formation of PhCH_2· from PhCH_2CHEtOH, cf. 2.2 for PhCH_2CH_2OH)[50] presumably reflect the side chain structures and, in particular, the increased stability of PhĊHMe and $^+$CH(OH)Et, respectively. Using a steady state approach, the rate constants for fragmentation have been estimated as 1.5×10^6 s^{-1} for [PhCH$_2$CH$_2$OH]‡, 2.5×10^6 s^{-1} for [PhCHMeCH$_2$OH]‡, and 1.5×10^7 s^{-1} for [PhCH$_2$CHEtOH]‡.[50] The radical cation from indan-2-ol (54) has been shown to give 55 on reaction with SO$_4^-$ below pH 4.5 and HO· below pH circa 4.[50] These high pH values suggest a rate constant for fragmentation of circa 3×10^7 s^{-1}; this high value presumably reflects the loss of ring strain and the stability of the resulting fragments. It also suggests that overlap between the orbital containing the unpaired electron and the C$_\alpha$–C$_\beta$ bond is not, as might have been expected, of paramount importance in determining the rate of this process.[50]

(54)   (55)

7.3. Decarboxylation

Reaction of Ph(CH$_2$)$_2$CO$_2$H and Ph(CH$_2$)$_3$CO$_2$H with HO· at pH values greater than 3.5 gives, as expected, ESR signals from a number of hydroxycyclohexadienyl radicals.[48,49,53] However, as the pH is lowered below this value signals from decarboxylated species (56, n = 1,2; Scheme 12) are observed and these in turn are replaced below pH circa 1.8 with those from the corresponding benzylic radicals (57, n = 1,2; Scheme 12)[48,49,53] Reaction with SO$_4^-$ in place of HO· gives rise to decarboxylated radicals at pH values > 1.5; below this value the signals from the benzylic radicals again dominate. In contrast, reaction of Ph(CH$_2$)$_4$CO$_2$H with HO· gives solely the hydroxycyclohexadienyl radicals above pH circa 1.5 and the corresponding benzylic radical below this value.[49,53]

The decarboxylated radicals are believed to arise via *overall* electron transfer from the carboxyl group to the arene radical cation (generated directly in the case of[38,54] SO$_4^-$ and via acid-catalyzed loss of H$_2$O from the hydroxyl adducts in the case of the attack of HO·[48,49,53]), resulting in the generation of arene carbonyloxy radicals Ph(CH$_2$)$_{n+1}$CO$_2$·, which are believed to undergo very ready loss of CO$_2$ (see Scheme 12).[55] The intramolecular process may involve either (or both) *direct* electron transfer (step a) or formation and rapid reopening of a discrete σ-bonded intermediate[49,53] (step b; cf. the reactions of the corresponding aliphatic alkeneoic acids mentioned earlier[41,47]).

Studies of examples in which the carboxyl group is held at some distance from the arene indicate that decarboxylation from the carboxylate anion still

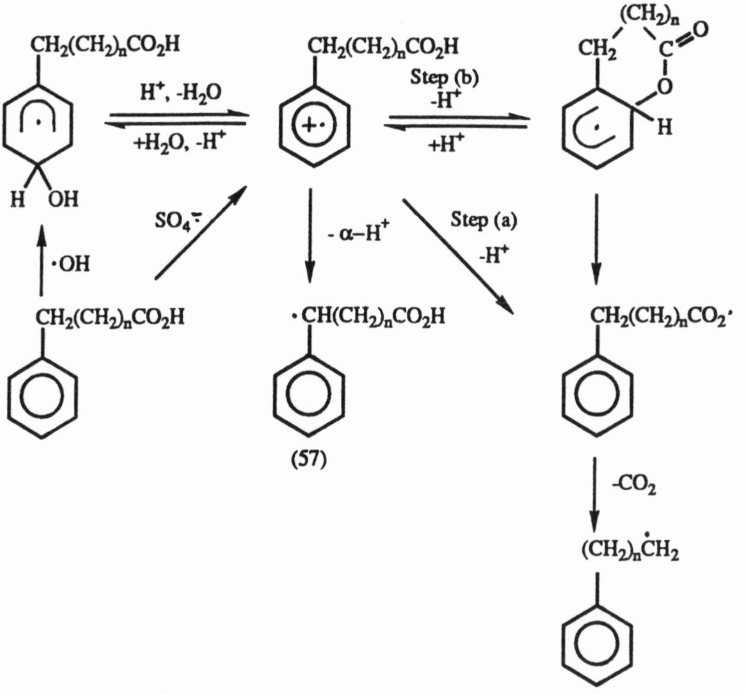

Scheme 12

occurs rapidly, i.e., that electron transfer from these anions is direct (through space) and rapid. *Protonation* of the carboxylate evidently retards this reaction, so that deprotonation (from α CH) is now preferred.

Reaction of HO· with the substrates $PhCH_2CH(OH)CO_2H$ and $PhCH_2CH(NH_3^+)COO^-$ allows a comparison to be made between the relative rates of decarboxylation (cf. reaction of $PhCH_2CH_2CO_2H$) and $C_\alpha–C_\beta$ cleavage (cf. $PhCH_2CH_2OH$). Reaction of HO· under acidic conditions (pH < 3) with the former substrate gives signals from $PhCH_2\overset{\cdot}{C}HOH$ together with traces of the benzyl radical (which grow in intensity as the pH is lowered further)[50] while with the latter substrate signals assigned to the radical adduct of $PhCH_2\overset{\cdot}{C}HNH_3^+$ have been observed in spin-trapping experiments using 2-methyl-2-nitrosopropane.[56] These results demonstrate that decarboxylation from the anion is faster than $C_\alpha–C_\beta$ cleavage, at least for the former substrate, but that protonation of the acid group reduces the rate of decarboxylation to a value less than that for cleavage (for which k_f has been estimated as $\approx 10^6$ s^{-1}).[50]

7.4. Remote Side Chain Deprotonation

Behavior analogous to that observed with the phenyl-substituted carboxylic acids is observed with certain phenyl-substituted alcohols. Thus reaction of HO· with $Ph(CH_2)_3OH$ gives hydroxycyclohexadienyl radicals above pH 3; below this value signals from the α-hydroxy-conjugated radical **58** (see Scheme 13) are observed and these in turn are replaced by those of the benzylic radical **59** at pH < 1.5.[50,53] As expected, increasing the length of the alkane chain [for example $Ph(CH_2)_4OH$] or incorporating the oxygen into an ether function [for example $Ph(CH_2)_3OMe$] gives signals only from the hydroxycyclohexadienyl radical (at pH > 1.5) and the benzylic radical (at pH < 1.5).[53] The formation of the α-hydroxy-conjugated species is believed to arise via formation of the radical **60** (see Scheme 13) followed by a rapid 1,2-hydrogen shift, which has been established for some related alkoxyl radicals under similar conditions.[50,53] As noted for the corresponding reactions of acids described earlier, this alkoxyl radical could be formed by either *direct* electron transfer from the alcohol function to the arene radical cation followed by rapid deprotonation or via the formation of a cyclic radical such as **61** (see Scheme 13). Direct electron transfer can probably be ruled out since such a process would be expected to occur more readily for the ether than the corresponding alcohol and in both cases would be expected to be unfavorable (cf. ionization potentials for toluene, diethyl ether, and ethanol of 8.82, 9.53, and 10.48 eV, respectively[46]). The formation in this case of the cyclic radical species (in a reaction that is rapidly reversed at very low pH) is in accord with the observation that such remote deprotonation processes do not occur when ring formation is prevented by either a longer side chain or the presence of the ether function,[53] and with the

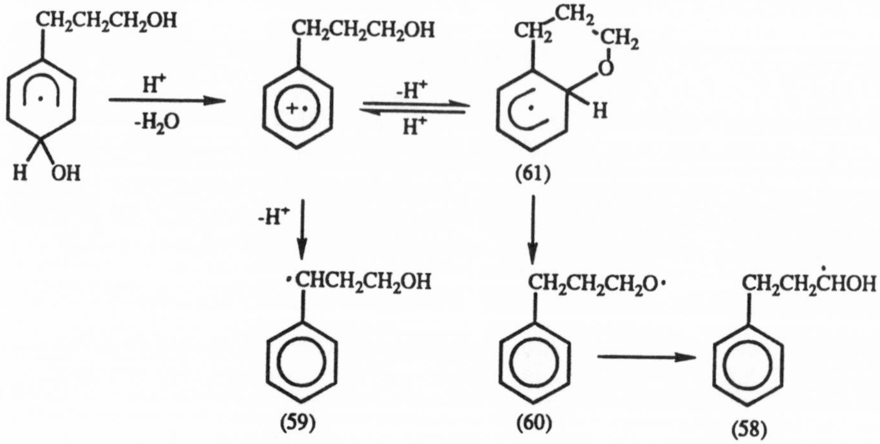

Scheme 13

(62)

isolation of cyclic products such as 3,4-dihydro-2*H*–1-benzopyran **(62)** from the reaction of HO· with Ph(CH₂)₃OH.[57]

7.5. Remote Side Chain Fragmentation

As might be expected if alkoxyl radicals are generated via the above processes, the presence of additional alkyl substituents can also result in remote side chain fragmentation. Thus reaction of HO· with both 2-methyl-4-phenyl-butan-2-ol (PhCH₂CH₂CMe₂OH) and 4-phenylbutan-2-ol (PhCH₂CH₂CHMeOH) gives rise to PhCH₂CH₂·, characteristic of the β fragmentation of an intermediate alkoxyl radical, together with PhCH₂CH₂ĊMeOH from the latter. The precursor alkoxyl species must arise via an intermediate from which the radical cation can be reformed in an acid-catalyzed process, since deprotonation at the α carbon (to give, e.g., PhĊHCH₂CMeOH) is observed at low pH[50]; these results are again interpreted in terms of the formation of a cyclic species via internal nucleophilic attack by the alcohol group on the radical cation (cf. Scheme 13).

8. RADICAL CATIONS FROM HETEROAROMATIC COMPOUNDS

Many of the processes outlined above for aromatic radical cations are also observed, as expected, with some simple heteroaromatic compounds such as thiophen and furan. With both of these compounds the initial attack of HO· at pH 4.5 can be shown, by use of either continuous-flow or in situ radiolysis ESR techniques, to give adduct species[58–60]; thus with thiophen the adducts at the 2 and 3 positions are observed in the ratio 4 : 1.[60] Reduction of the pH results in an increase in this ratio until by pH 2.5 the former radical is the only adduct observed. This species is also the only adduct observed on reaction of Cl₂⁻ with this substrate. This changeover can be explained in terms of the formation and (re)hydration of the corresponding radical cation; the predominance of the adduct at the 2 position presumably reflects the greater thermodynamic stability of the allylic system compared to the species produced by addition at the 3 position.[60]

Further reactions analogous to the aromatic reactions mentioned earlier are observed when substituents are introduced into the ring. Thus with 2-methyl-thiophen deprotonation at the side chain occurs in addition to hydration of the radical cation at low pH; the observation of both the deprotonated radical and hydrated species at very low pH with either HO· or Cl₂⁻ suggests that these two processes have similar rates.[60] With 2-methylfuran, however, the situation is somewhat different in that only hydration of the radical cation is observed (to give hydroxyl substituents at the 2 and 5 positions)[58]; this change in behavior has been ascribed to the greater stability of the 2-methylfuran radical cation (when compared to the 2-methylthiophen species) on account of the greater +M effect of the oxygen, as well as the extra stabilization afforded to the deprotonated species in the 2-methylthiophen case by spin delocalization onto sulfur.[60]

Oxidation of 2-thenylacetic acid at pH 4 gives, in addition to the expected hydroxyl radical adducts, signals from the 2-thenyl radical, which presumably arises from decarboxylation [see reaction (51)][60]; the concentration of the decarboxylated species increases dramatically at lower pH values and is the only radical observed at pH 0.5. This latter radical is the sole species observed on reaction with Cl_2^{-}. These results suggest that the radical cation, once formed, undergoes decarboxylation exclusively.[60]

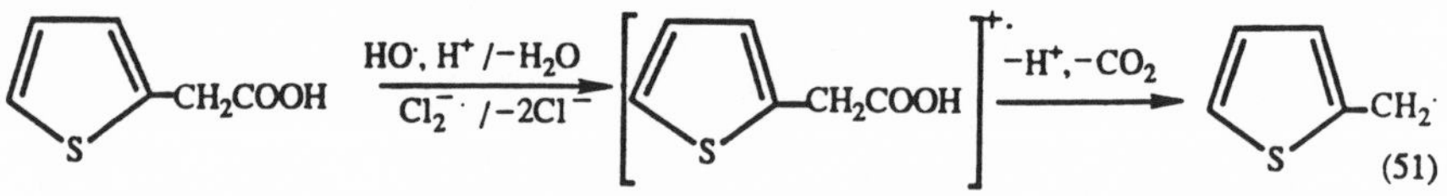

$$\text{(51)}$$

9. SULFUR-CENTERED RADICAL CATIONS

The oxidation of a number of sulfides by HO· and other powerful one-electron oxidants is found to be in many ways analogous to the oxidation processes described above for alkenes and aromatic compounds. Thus reaction of HO· with simple sulfides is believed to occur primarily via addition at sulfur to give R_2SOH; the formation of these adducts has not been directly observed in either ESR or pulse radiolysis experiments (owing to their short lifetimes and the fact that they do not exhibit any absorption in the visible and near-UV region[61]), but is in agreement with both the subsequently observed species and theoretical calculations. These species can evidently undergo two competing processes— either rapid loss of HO$^-$ to give[61-64] the monomeric radical cation R_2S^+ or

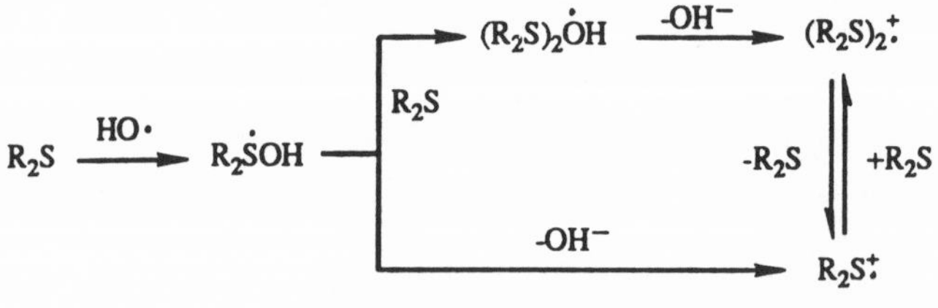

Scheme 14

reaction with the parent molecule to give a complexed species believed to be $(R_2S)_2\overset{.}{O}H$ (see Scheme 14)[63-65]; the latter process is facilitated at high sulfide concentrations. Examples of the former species have not been observed in either pulse radiolysis or ESR experiments involving ·OH at room temperature, but they are readily observed in frozen matrices.[66,67] These monomeric radical cations, unlike their alkene and arene counterparts, undergo very ready addition reactions (with parent) to give "dimeric" radical cations $R_2SSR_2^{\ddagger}$, which contain a three-electron bond between the two sulfur atoms; these species may also arise directly from decay of the $(R_2S)_2\overset{.}{O}H$ complexes.[63] The dimeric radical cations have been observed directly in both ESR continuous-flow experiments at low pH[62] and in pulse–radiolysis studies[63,65]: the proton hyperfine splittings in their ESR spectra are much smaller (by a factor of circa 3) than those of the corresponding monomer radical cations [cf. values of 2.04 and 0.68 mT for the protons on $Me_2S^{\ddagger}$ and $(Me_2S)_2^{\ddagger}$, respectively[62,66]] and it is concluded that the SOMO in the dimer is a σ^* molecular orbital. Similar behavior is observed with the analogous selenium species.[66]

The equilibrium between the monomeric and dimeric radical cations has been investigated in pulse–radiolysis experiments via observation of the decay of the absorption of the dimeric species (at circa 470 nm[63]); the decay process is complex and has been analysed in terms of first-order rate constants for the decay of the monomer radical cation $R_2S^{\ddagger}$ and the equilibrium constant for the dissociation of the dimer. A variety of subsequent reactions of the monomeric species $R_2S^{\ddagger}$ formed either via dissociation or from direct acid-catalyzed loss of HO^- from the initial $R_2S\overset{.}{O}H$ adduct, have been revealed by ESR and pulse–radiolysis studies. In many ways they are analogous to those of the alkene and arene radical cations described earlier.

For example, *loss of an α proton* to give a stabilized sulfur-conjugated radical has been observed by ESR for a large number of dialkyl sulfides on reaction with HO· [reaction (52)][63]; this appears to be the major pathway at low dialkyl sulfide concentrations and at high pH values. Thus, while oxidation of dimethyl sulfide with the HO· radical at pH 2.5 gives strong ESR signals from the dimeric radical cation $(Me_2S)_2^{\ddagger}$, at pH 3.6 only the deprotonated radical ·CH$_2$SMe is observed.[62] A second pathway for formation of ·CH$_2$SMe, of relatively minor importance, is direct attack of HO· at the α carbon [reaction (53)]; this type of pathway has been estimated from pulse–radiolysis studies[65] to account for less than 20% of the total flux of HO· reacting with, for example, methionine (i.e., there is >80% addition at the sulfur center). The third pathway, which has also been elucidated by pulse–radiolysis experiments, is rapid elimination of H_2O ($k > 7 \times 10^5$ s^{-1} for diethyl sulfide[63]) from the initially generated $R_2\overset{.}{S}OH$ radicals [reaction (54)].

The detection of thiyl radials by ESR spin trapping using the aci-anion of nitromethane [reaction (55)] during the oxidation of certain β hydroxy sulfides with HO· at pH 9 is consistent with *heterolytic carbon–sulfur bond cleavage* in

$$\overset{+\cdot}{MeSCH_2CH_2OH} \quad \xrightarrow{-H^+} \quad \cdot CH_2SCH_2CH_2OH \qquad (52)$$

$$MeSCH_2CH_2C(NH_3^+)COO^- \quad \xrightarrow{HO\cdot} \quad \begin{cases} \cdot CH_2SCH_2CH_2C(NH_3^+)COO^- \\ \\ Me\overset{\cdot}{S}CHCH_2C(NH_3^+)COO^- \end{cases} \qquad (53)$$

$$MeCH_2SCH_2Me \quad \xrightarrow{HO\cdot} \quad \underset{\underset{OH}{|}}{MeCH_2\overset{\cdot}{S}CH_2Me} \quad \longrightarrow \quad Me\overset{\cdot}{C}HSCH_2Me + H_2O \qquad (54)$$

the monomeric sulfur radical cation [reaction (56)].[61] The detection of such radicals is also consistent with direct attack of HO· at the β carbon positions and homolytic cleavage [e.g., reaction (57)]. However, the detection of thiyl radicals from [Me$_2$C(OH)CH$_2$]$_2$S, which cannot undergo this latter process, suggests that at least in some cases the former mechanism is dominant.[61] The detection by ESR of signals due to ·CH$_2$SCH$_2$Me on oxidation of HOCH$_2$CH$_2$SCH$_2$Me with HO· at pH 4.5 suggests that C_α–C_β *bond cleavage* is also a relatively ready process [reaction (58)].[61] The observation of this species (and related radicals with other substrates[61]) is directly analogous to the

$$RS\cdot + CH_2=NO_2^- \quad \longrightarrow \quad RSCH_2NO_2^{-\cdot} \qquad (55)$$

$$HOCH_2CH_2SCH_2CH_2OH \quad \xrightarrow{HO\cdot} \quad HO-\overset{H}{\underset{}{\overset{|}{C}H}}\!-\!CH_2\!-\!\overset{+\cdot}{S}CH_2CH_2OH \quad \longrightarrow \quad \begin{matrix} HOCH=CH \\ + \\ \cdot SCH_2CH_2OH + H^+ \end{matrix} \qquad (56)$$

$$R'CH(OH)CH_2SR \quad \xrightarrow{HO\cdot} \quad R'\overset{\cdot}{C}(OH)CH_2SR \quad \longrightarrow \quad R'C(OH)=CH_2 + \cdot SR \qquad (57)$$

observation of benzyl radicals on oxidation of β-phenylethanol with HO· described earlier (see Scheme 11). This process appears, at least with HOCH$_2$CH$_2$SCH$_2$Me, to occur at approximately the same rate as deprotonation from the α carbon positions, since the radicals ·CHMeSCH$_2$CH$_2$OH and ·CH$_2$SCH$_2$Me are observed in the ratio 3 : 1.[61] Similar cleavage processes are believed to underlie the formation of the radical ·CH$_2$SCH$_2$CO$_2$H from HO$_2$CCH$_2$SCH$_2$CO$_2$H on reaction with HO· at pH 4.5 [reaction (59)].[61] In this latter case the radical HO$_2$ĊHSCH$_2$CO$_2$H, which would arise from deprotona-

$$HOCH_2CH_2SCH_2CH_2OH \quad \xrightarrow{HO\cdot} \quad H\overset{\frown}{-O}\!-\!CH_2\overset{\frown}{-}CH_2\overset{+\cdot}{S}CH_2CH_2OH \qquad (58)$$

$$\downarrow$$

$$\cdot CH_2SCH_2CH_2OH + H^+ + CH_2=O$$

$$HOOCCH_2\overset{+\cdot}{S}CH_2COOH \quad \longrightarrow \quad H^+ + CO_2 + \cdot CH_2SCH_2COOH \qquad (59)$$

tion at the α carbon positions, is not observed, suggesting that decarboxylation (as with the analogous aromatic reactions, see earlier) is rapid compared to deprotonation.[61]

Remote *side chain decarboxylation* reactions analogous to the processes described earlier for $Ph(CH_2)_2COOH$ and $Ph(CH_2)_3COOH$ do not appear to occur as readily with the analogous monomeric sulfur radical cations. Thus, oxidation of $EtS(CH_2)_3CO_2^-$ with HO· gives rise solely to signals from ·$CHMeS(CH_2)_3COO^-$ and ·$CH(SEt)CH_2CH_2CO_2^-$ and not ·$CH_2CH_2CH_2SEt$, which would be formed if oxidation of the carboxylate ion had occurred either directly or through intramolecular electron transfer to the sulfur center in the radical cation.[68] However, the introduction of groups onto the carbon atom adjacent to the carboxylate function, which would be expected to stabilize a decarboxylated radical, appears to make this a viable process. Thus, oxidation of methionine [$MeSCH_2CH_2CH(CO_2^-)NH_3^+$], S-methylcysteine [$MeSCH_2CH(CO_2^-)NH_3^+$], and 2-hydroxy-4-(methylthio)butanoic acid [$MeSCH_2CH_2CH(CO_2^-)OH$] with HO· at pH values between 2.5 and 4.5 gives rise to the detection by ESR spectroscopy of the radicals $MeSCH_2CH_2\dot{C}HNH_3^+$ (trapped with *tert*-BuNO), $MeSCH_2\dot{C}HNH_3^+$ (trapped with *tert*-BuNO), and $MeSCH_2CH_2\dot{C}HOH$ (trapped with $MeNO_2$ and *tert*-BuNO), respectively.[68] At pH values > 4.5 the corresponding unprotonated α-aminoalkyl radical is trapped with *tert*-BuNO in the case of methionine.[68] Complementary pulse–radiolysis experiments[65] have conclusively demonstrated the generation of α-aminoalkyl radicals in the ·OH-induced oxidation of methionine.

The mechanism of this side chain oxidation has been discussed in some detail.[68] It has been clearly demonstrated by pulse–radiolysis that overall one-electron transfer occurs *from* the carboxylate function *to* the precursor

$$Me\overset{+\cdot}{S}CH_2CH_2CH\overset{NH_3^+}{\underset{COO^-}{\diagdown}} \longrightarrow MeSCH_2CH_2CH\overset{NH_3^+}{\underset{COO^\cdot}{\diagdown}} \xrightarrow{-CO_2} MeSCH_2CH_2\dot{C}H-NH_3^+ \quad (60)$$

$$Me\overset{+\cdot}{S}CH_2CH_2CH\overset{NH_3^+}{\underset{COO^-}{\diagdown}} \longrightarrow \underset{(63)}{\text{[cyclic intermediate]}} \xrightarrow{-CO_2} MeSCH_2CH_2\dot{C}H-NH_3^+ \quad (61)$$

(63)

monomeric sulfur radical cation, resulting in the formation of CO_2 and an α-aminoalkyl radical.[65] However, two different mechanisms have been proposed for the mechanism of transfer. ESR spin-trapping studies[68] have been interpreted in terms of either direct transfer from the carboxylate to the radical cation center [reaction (60)] or the formation of a cyclic intermediate [such as

63 in reaction (61); cf. reactions of alkene and aromatic carboxylic acids mentioned earlier].

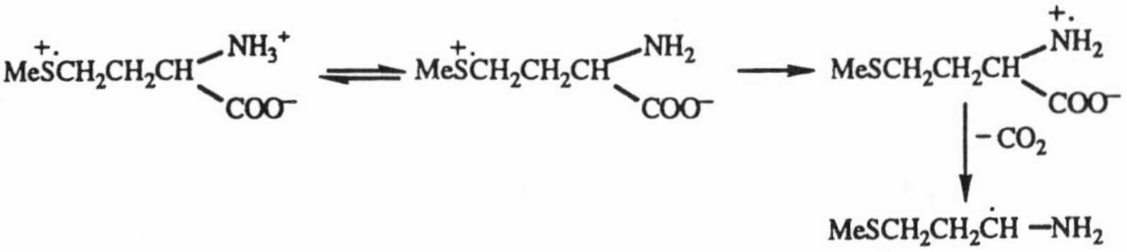

$$\text{(62)}$$

In contrast, pulse–radiolysis data[65] have been interpreted in terms of electron transfer from the *amino* group to the sulfur radical cation followed by interaction of the aminium radical cation with the carboxyl group to give CO_2 and the α-aminoalkyl radical; such a process [reaction (62)] would be expected to be sterically assisted as the participating sulfur and nitrogen centers can adopt a favorable five-membered ring configuration (**64**).[65]

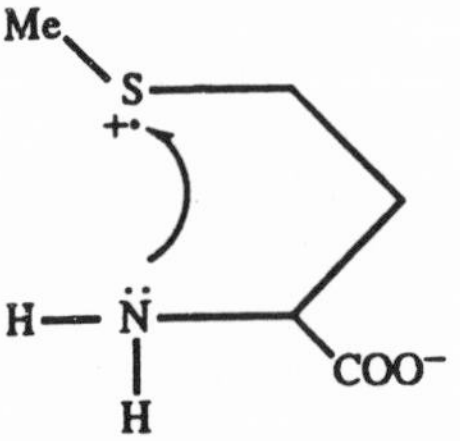

(64)

Direct evidence for the oxidation of amine groups by sulfur radical cations has been obtained by pulse–radiolysis[65] and it would appear, at least for

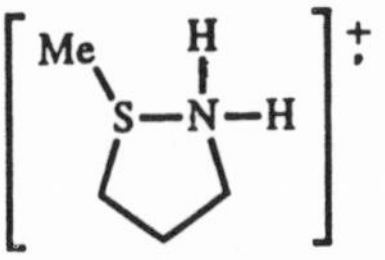

(65) 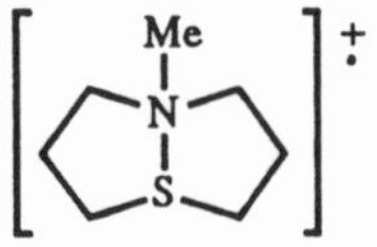   (66)

methionine, that the latter mechanism is favored. Further experimental evidence has been obtained for the mediation of cyclic species such as **63** and **64** in electron transfer processes. Thus, 3-electron 2-center cyclic species such as **65** and **66** have been observed on oxidation of 3-(methylthio)propylamine[69] and 5-methyl-l-thia-5-azacyclooctane,[70] respectively, and cyclic species such as **68** [reaction (63)] and **70** [reaction (64)] have been observed on laser flash photolysis of **67**[71] and pulse radiolysis of **69**,[72] respectively.

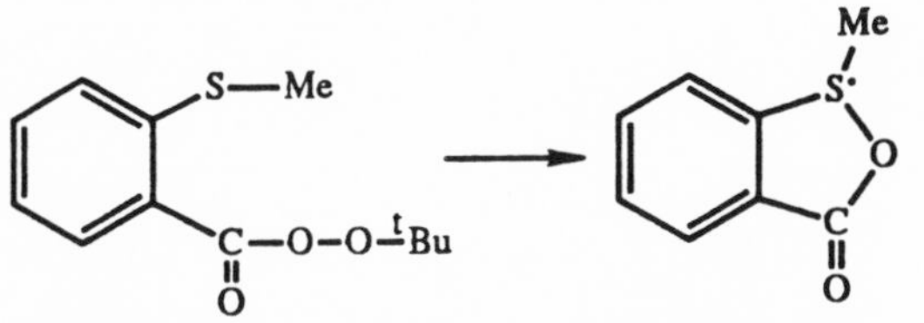

(63)

(67)   (68)

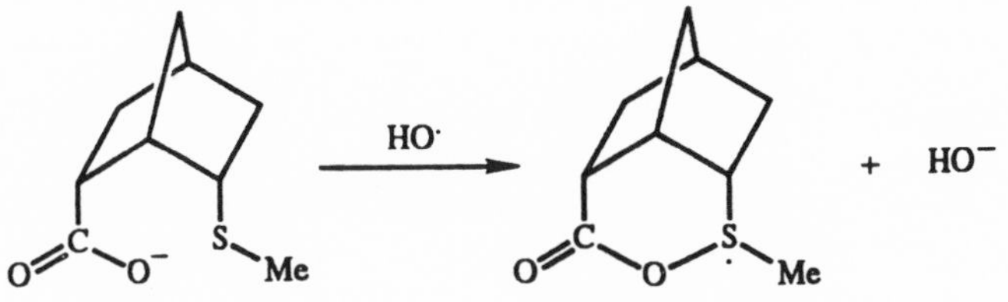

(64)

(69)   (70)

ACKNOWLEDGMENTS

In our quest for direct spectroscopic evidence for novel reactive intermediates—especially radical cations—and for unified features of acid- and base-catalyzed reactions that link the free-radical chemistry of alkenes, arenes, and sulfur compounds, we have collaborated with a number of colleagues (mainly pre- and postdoctoral research students) at the University of York. To these, coauthors in the papers to which we have referred, we express our gratitude for their skills, enthusiasm, and stimulating ideas. It is also a pleasure to thank Dr. Steen Steenken and his colleagues at the Max-Planck-Institut für Strahlenchemie, Mülheim, Germany, for sharing ideas, for helpful discussions, and for experimental assistance. Financial support from the SERC is gratefully acknowledged.

REFERENCES

1. Giese, B. *Radicals in Organic Synthesis: Formation of Carbon–Carbon Bonds;* Pergamon Press: Oxford, 1986.
2. Nonhebel, D. C.; Tedder, J. M.; Walton, J. C. *Radicals*; Cambridge University Press: Cambridge, 1979.
3. Halliwell, B.; Gutteridge, J. M. C. *Free Radicals in Biology and Medicine*, Clarendon Press: Oxford, 1985.
4. Ingold, K. U.; Beckwith, A. L. J. In *Rearrangements in Ground and Excited States*; de Mayo, P., Ed.; Academic Press: New York, 1980, Vol. 1, p 161.
5. von Sonntag, C. *The Chemical Basis of Radiation Biology*; Taylor and Francis; London, 1987.
6. von Sonntag, C. *Adv. Carbohydr. Chem. Biochem.* **1980**, *37*, 7.
7. (a) Adams, G. E.; Wardman, P. In *Free Radicals in Biology*; Pryor, W. A., Ed; Academic Press: New York, 1977, Vol 3, p 1. (b) Willson, R. L.; Greenstock, C. L.; Adams, G. E.; Wageman, R.; Dorfman, L. M. *Int. J. Radiat. Phys. Chem.* **1971**, *3*, 211.

8. Norman, R. O. C.; Gilbert, B. C. *Adv. Phys. Org. Chem.* **1967**, *5*, 53.
9. Norman, R. O. C. *Chem. Soc. Rev.* **1979**, *8*, 1.
10. Kochi, J. K. *Adv. Free-Radical Chem.* **1975**, *5*, 189.
11. (a) Buley, A. L.; Norman, R. O. C.; Pritchett, R. J. *J. Chem. Soc. B* **1966**, 849. (b) Livingston, R.; Zeldes, H. *J. Am. Chem. Soc.* **1966**, *88*, 4333.
12. Bansal, K. M.; Grätzel, M.; Henglein, A.; Janata, E. *J. Phys. Chem.* **1973**, *77*, 16.
13. Steenken, S.; Davies, M. J.; Gilbert, B. C. *J. Chem. Soc., Perkin Trans. 2* **1986**, 1003.
14. (a) Dobbs, A. J.; Gilbert, B. C.; Norman, R. O. C. *J. Chem. Soc., Perkin Trans. 2*, **1972**, 786. (b) Gilbert, B. C.; Norman, R. O. C.; Dobbs, A. J. *J. Mag. Resonance* **1973**, *11*, 100.
15. Steenken, S. *J. Phys. Chem.* **1979**, *83*, 595.
16. Gilbert, B. C.; King, D. M.; Thomas, C. B. *J. Chem. Soc., Perkin Trans. 2* **1980**, 1821.
17. Gilbert, B. C.; King, D. M.; Thomas, C. B. *J. Chem. Soc., Perkin Trans. 2* **1981**, 1186.
18. Gilbert, B. C.; King, D. M.; Thomas, C. B. *Carbohydr. Res.* **1984**, *125*, 217.
19. Gilbert, B. C.; King, D. M.; Thomas, C. B. *J. Chem. Soc., Perkin Trans. 2* **1983**, 675.
20. Gilbert, B. C.; Larkin, J. P.; Norman, R. O. C. *J. Chem. Soc., Perkin Trans. 2* **1972**, 794.
21. Foster, T.; West, P. R. *Can. J. Chem.* **1974**, *52*, 3589.
22. Behrens, G.; Koltzenburg, G.; Schulte-Frohlinde, D. *Z. Naturforsch.* **1982**, *37c*, 1205.
23. (a) Behrens, G.; Bothe, E.; Eibenberger, J.; Koltzenburg, G.; Schulte-Frohlinde, D., *Int. J. Radiat. Biol.* **1978**, *33*, 163. (b) Behrens, G.; Bothe, E.; Koltzenburg, G. and Schulte-Frohlinde, D. *J. Chem. Soc., Perkin Trans. 2* **1980**, 883.
24. Behrens, G.; Bothe, E.; Koltzenburg, G.; Schulte-Frohlinde, D. *J. Chem. Soc., Perkin Trans. 2* **1981**, 143.
25. Behrens, G.; Koltzenburg, G, *Z. Naturforsch.* **1985**, *40c*, 785.
26. Samuni, A.; Neta, P. *J. Phys. Chem.* **1973**, *77*, 2425.
27. Fitchett, M.; Gilbert, B. C.; Willson, R. L. *J. Chem. Soc., Perkin Trans. 2* **1988**, 673.
28. Behrens, G.; Koltzenburg, G.; Ritter A.; Schulte-Frohlinde, D. *Int. J. Radiat. Biol.* **1978**, *33*, 163.
29. (a) Deeble, D. J.; von Sonntag, C. *Int. J. Radiat. Biol.* **1984**, *46*, 247. (b) Schulte-Frohlinde, D.; Bothe, E. *Z. Naturforsch.* **1984**, *39c*, 315. (c) Hildebrand, K.; Behrens, G.; Schulte-Frohlinde, D.; Herak, J. N. *J. Chem. Soc. Perkin Trans. 2* **1989**, 293. (d) Schulte-Frohlinde, D.; Hildebrand, K. In *Free Radicals in Synthesis and Biology*; F. Minisci, Ed.; NATO ASI, 1989, Kluwer, Dordrecht, Vol 260, p 335.
30. Burkitt, M. J.; Fitchett, M.; Gilbert, B. C. In *Medical, Biochemical and Chemical Aspects of Free Radicals*; O. Hagaishi, E. Niki, M. Kondo, T. Yoshikawa, Eds.; Elsevier: Amsterdam, 1989, p 63.
31. Beckwith, A. L. J.; Tindal, P. K. *Aust J. Chem.* **1971**, *24*, 2099.
32. Gilbert, B. C.; Norman, R. O. C.; Williams, P. S. *J. Chem. Soc., Perkin Trans. 2* **1981**, 1401.
33. Golding, B. T.; Radom, L. *J. Am. Chem. Soc.* **1976**, *98*, 6331.
34. Asmus, K.-D.; Williams, P. S.; Gilbert, B. C.; Winter, J. N. *J. Chem. Soc., Chem. Commun.* **1987**, 208.
35. Beckwith, A. L. J.; Duggan, P. J. *J. Chem. Soc., Chem. Commun.* **1988**, 1000.
36. Kocovsky, P.; Stary, L; Turecek, F. *Tetrahedron Lett.* **1986**, *27*, 1513.
37. Barclay, L. R. C.; Lusztyk, J.; Ingold, K. U. *J Am. Chem. Soc.* **1984**, *106*, 1793.
38. Norman, R. O. C.; Storey, P. M.; West, P. R. *J. Chem. Soc. B* **1970**, 1087.
39. Davies, M. J.; Gilbert, B. C.; Norman, R. O. C. *J. Chem. Soc., Perkin Trans. 2* **1984**, 503.
40. (a) Hasegawa, K.; Neta, P. *J. Phys. Chem.* **1978**, *82*, 854. (b) Jayson, G. G.; Parsons, B. J.; Swallow, A. J. *J. Chem. Soc., Faraday Trans. 1* **1973**, 1597.
41. Davies, M. J.; Gilbert, B. C. *J. Chem. Soc., Perkin Trans. 2* **1984**, 1809.
42. Koltzenburg, G.; Behrens, G.; Schulte-Frohlinde, D. *J. Am. Chem. Soc.* **1982**, *104*, 7311.
43. Koltzenburg, G.; Bastian, E.; Steenken, S. *Angew. Chem., Int. Ed. Engl.* **1988**, *27*, 1066.

44. See, e.g., Beckwith, A. L. J.; Easton, C. J.; Serelis, A. K. *J. Chem. Soc., Chem. Commun.* **1980**, 482, and references therein.
45. (a) Eibenberger, H.; Steenken, S.; O'Neill, P.; Schulte-Frohlinde, D. *J. Phys. Chem.* **1978**, 82, 749. (b) Clerici, A.; Minisci, F.; Ogawa, K.; Surzur, J.-M., *Tetrahedron Lett.* **1978**, 1149.
46. See, e.g., Kiser, R. W. *Introduction to Mass Spectrometry and its Applications*; Prentice-Hall: Englewood Cliffs, 1965, pp 308–320.
47. Davies, M. J.; Gilbert, B. C. *J. Chem. Res. (S)* **1985**, 162.
48. (a) Norman, R. O. C.; Pritchett, R. J. *J. Chem. Soc. B* **1967**, 926. (b) Norman, R. O. C.; Storey, P. M. *J. Chem. Soc. B* **1970**, 1099.
49. Gilbert, B. C.; Scarratt, C. J.; Thomas, C. B.; Young, J. *J. Chem. Soc., Perkin Trans.* 2 **1987**, 2, 371.
50. Gilbert, B. C.; Warren, C. J. *Res. Chem. Int.* **1989**, 11, 1.
51. (a) Walling, C.; El-Taliawi, G. M.; Zhao, C., *J. Org. Chem.,* **1983**, 48, 4914. (b) Walling, C.; Zhao, C.; El-Taliawi, G. M. *J. Org. Chem.,* **1983**, 48, 4910. (c) Snook, M. E.; Hamilton, G. A. *J. Am. Chem. Soc.,* **1974**, 96, 860.
52. Walling, C.; Camaioni, D. M.; Kim, S. S. *J. Am. Chem. Soc.,* **1978**, 100, 4814.
53. Davies, M. J.; Gilbert, B. C.; McCleland, C. W.; Thomas, C. B.; Young, J., *J. Chem. Soc., Chem. Commun.* **1984**, 966.
54. (a) Zemel, H.; Fessenden, R. W. *J. Phys. Chem.,* **1975**, 79, 1419. (b) Madhavan, V.; Levanon, H.; Neta, P. *Radiat. Res.,* **1978**, 76, 15.
55. See, e.g., Clerici, A.; Minisci, F.; and Porta, O. *Tetrahedron Lett.* **1974**, 4183; Walling, C.; Camaioni, D. M. *J. Org. Chem.,* **1978**, 43, 3266; Giordano, C.; Belli, A.; Citterio, A.; Minisci, F. *J. Chem. Soc., Perkin Trans. 1* **1981**, 1574.
56. Fitchett, M.; Gilbert, B. C.; Jeff, M. *Phil. Trans. R. Soc. (London)* **1985**, B311, 517.
57. (a) Taylor, E. C.; Andrade, J. G.; Rall, G. J. H.; Turchi, I. J.; Steliou, K.; Jagdmann, G. E.; McKillop, A. *J. Am. Chem. Soc.* **1981**, 103, 6856. (b) Gilbert, B. C.; McCleland, C. W. *J. Chem. Soc., Perkin Trans.* 2 **1989**, 1545.
58. Schuler, R. H.; Laroff, G. P.; Fessenden, R. W. *J. Phys. Chem.* **1973**, 77, 456.
59. Gilbert, B. C.; Norman, R. O. C.; Williams, P. S. *J. Chem. Soc., Perkin Trans.* 2 **1980**, 647.
60. Gilbert, B. C.; Norman, R. O. C.; Williams, P. S. *J. Chem. Soc., Perkin Trans.* 2 **1981**, 207.
61. Gilbert, B. C.; Larkin, J. P.; Norman, R. O. C. *J. Chem. Soc., Perkin Trans.* 2 **1973**, 272.
62. Gilbert, B. C.; Hodgeman, D. K. C.; Norman, R. O. C. *J. Chem. Soc., Perkin Trans.* 2 **1973**, 1748.
63. Bonifačič, M.; Möckel, H.; Bahnemann, D.; Asmus, K.-D. *J. Chem. Soc., Perkin Trans.* 2 **1975**, 675.
64. Meissner, G.; Henglein, A.; Beck, G. *Z. Naturforsch.* **1967**, 22b, 13.
65. Hiller, K.-O.; Masloch, B.; Gobl, M.; Asmus, K.-D. *J. Am. Chem. Soc.* **1981**, 103, 2734.
66. Wang, J. T.; Williams, F. *J. Chem. Soc., Chem. Commun.* **1981**, 1184.
67. Rao, D. N. R.; Symons, M. C. R.; Wren, B. W. *J. Chem. Soc., Perkin Trans 2* **1984**, 1681.
68. Davies, M. J.; Gilbert, B. C.; Norman, R. O. C. *J. Chem. Soc., Perkin Trans.* 2 **1983**, 731.
69. Asmus, K.-D.; Gobl, M.; Hiller, K.-O.; Mahling, S.; Monig, J. *J. Chem. Soc., Perkin Trans.* 2 **1985**, 641.
70. (a) Musker, W. K.; Hirschon, A. S.; Doi, J. T. *J. Am. Chem. Soc.* **1976**, 100, 7754. (b) Musker, W. K.; Surdhar, P. S.; Ahmad, R.; Armstrong, D. A. *Can. J. Chem.* **1984**, 62, 1874.
71. Perkins, C. W.; Martin, J. C.; Arduengo, A. J.; Lau, W.; Alegria, A.; Kochi, J. K. *J. Am. Chem. Soc.* **1980**, 102, 7753.
72. Mahling, S.; Asmus, K.-D.; Glass, R. S.; Hojjatie, M.; Sabahi, M.; Wilson, G. S. *J. Org. Chem.* **1987**, 52, 3717.

α-CARBON-CENTERED RADICALS FROM AMINO ACIDS AND THEIR DERIVATIVES

Christopher J. Easton

Advances in Detailed Reaction Mechanisms
Volume 1, pages 83–126
Copyright © 1991 JAI Press Inc.
All rights of reproduction in any form reserved.
ISBN: 1-55938-164-7

83

1. INTRODUCTION

Many biosynthetic and biodegradative reactions of proteins, peptides, and other amino acid derivatives involve free-radical processes. For example, there is considerable evidence that free radicals are involved in penicillin and cephalosporin biosynthesis,[1–5] and they have been proposed as intermediates in protein–DNA crosslinking[6–8] and in the yellowing of protein.[9] Radical reactions of amino acids and their derivatives have been studied in order to investigate biochemical systems such as these[5,10,11] and to develop synthetic methods.[12–27] With over 500 amino acids now known, of which circa 240 occur free in nature,[28] these compounds and their derivatives are of interest as synthetic targets.

The free radicals that have been identified in proteins upon irradiation[29,30] may be categorized into three types: aromatic radicals, sulfur radicals, and aliphatic radicals. Aromatic radicals arise from reactions of aromatic side chains in amino acid residues such as phenylalanine and tryptophan.[31] The production of sulfur radicals is thought to involve sensitization of sulfur-containing side chains by aromatic residues.[32–34] The aliphatic radicals produced through the irradiation of proteins are mainly α-carbon-centered radicals (Figure 1). They have been chosen as the focus of this review because they are the radicals that are particular to amino acid derivatives. Other radicals are discussed only in terms of their reactions competing with those of α-carbon-centered radicals. Functional group manipulations such as deaminations and decarboxylations, in which the integrity of the amino acids and their derivatives is not retained, are not discussed. These and other aspects of the radical chemistry of amino acids and their derivatives have been the subject of previous reviews.[29,30]

The main aliphatic radicals produced in proteins upon irradiation have been identified as α-carbon-centered radicals derived from glycine residues.[30] Their electron spin resonance (ESR) spectra show a doublet resonance with hyperfine

$$\overset{+}{H_3N}-CH_2-CONH-\overset{\bullet}{C}H-CO_2^- \qquad MeCONH-\overset{\bullet}{C}H-CO_2^-$$
$$\qquad\qquad\qquad 1 \qquad\qquad\qquad\qquad\qquad\qquad 2$$

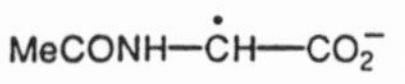

coupling similar to that observed with the radicals **1** and **2**, generated from glycylglycine and N-acetylglycine, respectively.[35–42] The spectra of **1** and **2** show that there is extensive conjugation in radicals of this type, with only 70–75% of the unpaired spin density on the α carbon.[30,35,36,42]

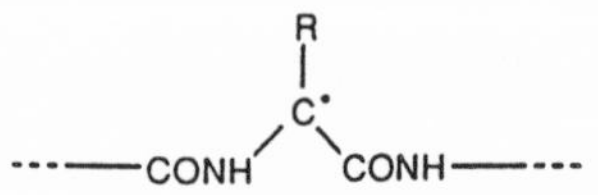

Figure 1. Aliphatic radicals formed by irradiation of proteins.

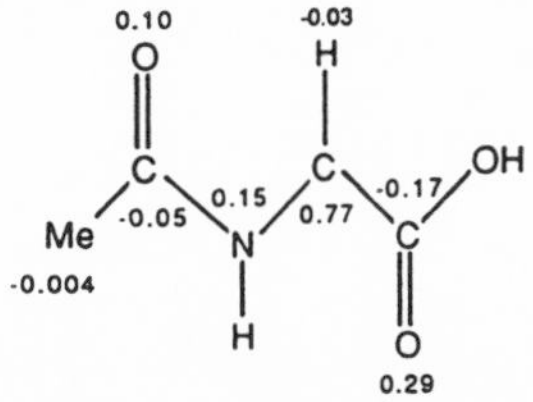

Figure 2. Distribution of the unpaired spin density in the radical produced by hydrogen atom transfer from N-acetylglycine.[30]

Molecular orbital calculations have also shown that the unpaired spin density in the radical 2 is distributed over the molecule (Figure 2).[30] The main contribution is from the α center with other contributions from both the carboxyl and amido groups. Radicals of this type belong to the class of captodative radicals. The captodative effect was postulated by Viehe et al.[43,44] as the combined resonance effect of electron-withdrawing (capto) and electron-donating (dative) substitutents on a radical center, leading to enhanced stabilization of the radical. The theoretical basis of this concept was originally formulated by Dewar in 1952.[45] Analogous concepts of "push–pull" radicals and "merostabilization" were independently developed by Balaban[46-48] and by Katritzky et al.,[49,50] respectively. Although the presence or absence of a synergistic radical-stabilizing effect, when a radical is substituted by both electron-donor and electron-acceptor moieties, remains a matter of debate,[51-59] the combined but not necessarily synergistic action of both the substituents results in stabilization of the radical.

Other amido- and carboxyl-substituted radicals analogous to 1 and 2 have been detected by electron spin resonance spectroscopy. The radicals 3 and 4 have been identified in studies of N-acetylalanine[37,40,41,60] and pyroglutamic acid,[60,61] respectively. With dipeptides, α-carbon-centered radicals are formed through reaction of C-terminal amino acids.[37,40-42,62] Thus glycylglycine and glycylalanine afford the radicals 1 and 5, respectively. The selective reaction of C-terminal amino acid residues in dipeptides can be attributed to the stability of the product radicals. For example, the radical 6 is considerably less stable than 1. The ammonium group strongly destabilizes a radical centered at the adjacent position.[39,41] For this reason α-carbon-centered radicals have been most frequently observed in reactions of amino acid derivatives, rather than with free amino acids. Only the nonprotonated radicals 7a, 7b, and 8 have been

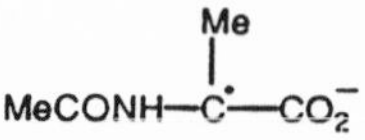

Me
|
MeCONH—C•—CO$_2^-$

3

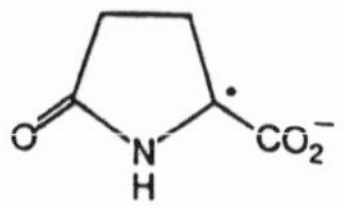

4

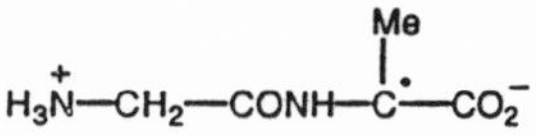

Me
|
$\overset{+}{H_3N}$—CH$_2$—CONH—C•—CO$_2^-$

5

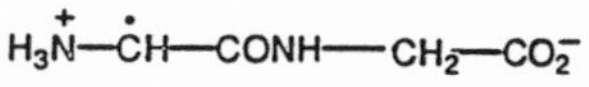

$\overset{+}{H_3N}$—•CH—CONH——CH$_2$—CO$_2^-$

6

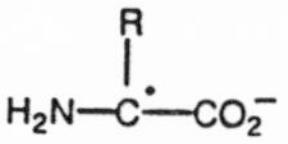

R
|
H$_2$N—C•—CO$_2^-$

7

H$_2$N—•CH—CH$_2$—CO$_2^-$

8

a) R = H
b) R = Me

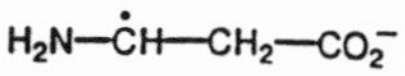

identified in ESR studies of glycine and α- and β-alanine, respectively.[39,63,64] The amino group stabilizes a radical formed at the adjacent position.

The numerous ESR studies have provided valuable qualitative data on the types of radicals produced through reaction of amino acids and their derivatives. The majority of the examples chosen to illustrate this chapter, however, are from studies for which quantitative data are available. Accordingly, the emphasis is on those investigations in which products have been isolated or characterized. Studies in which the conclusions are based solely on the spectroscopic observation of transient species, or on otherwise wholly qualitative data, are included only if they exemplify points raised in other investigations.

2. RADICALS FORMED BY HYDROGEN TRANSFER

Most of the studies in which products have been isolated have involved hydrogen-atom transfer reactions of amino acid derivatives. In a series of papers[65-71] Elad et al. have reported photoalkylation reactions of amino acid derivatives as a tool for the modification of peptides and proteins. In one of many examples,[67] irradiation with ultraviolet light of a mixture of the dipeptide derivative 9 and but-1-ene, in the presence of acetone at room temperature, gave the modified dipeptide derivatives 10a and 11a, in yields of 17% and 14%, respectively. With toluene instead of but-1-ene, 10b and 11b were formed in yields of 26% and 35%, respectively.[67]

A mechanism involving free-radical intermediates was proposed for the photoalkylation reactions (Scheme 1). Absorption of incident light by acetone

CF₃CONH—CH₂—CONH—CH₂—CO₂Me

9

$$\text{CF}_3\text{CONH}\overset{\displaystyle R}{\overset{|}{-}\text{CH}}\text{—CONH—CH}_2\text{—CO}_2\text{Me} \qquad\qquad \text{CF}_3\text{CONH—CH}_2\text{—CONH}\overset{\displaystyle R}{\overset{|}{-}\text{CH}}\text{—CO}_2\text{Me}$$

10   **11**

a) R = CH₂CH₂CH₂Me
b) R = CH₂Ph
c) R = CH₂CH(CH₂Me)CH₂CH₂CH₂Me

produces the triplet ketone, which abstracts a hydrogen atom from the amino acid derivative to give an α-carbon-centered radical. That radical reacts with but-1-ene by addition. In support of the proposed mechanism, low-molecular-weight telomers were identified in the photoalkylation reactions. For example, evidence was obtained for the production of **10c** and **11c** in the reaction of **9** with but-1-ene. For reactions carried out in the presence of toluene, the α-carbon-centered radical reacts by combination with benzyl radical, formed by hydrogen atom transfer from toluene. Bibenzyl was also formed in reactions when toluene was used, consistent with dimerization of benzyl radical. In the absence of an alkylating agent, dehydrodimers of the amino acid derivatives were formed, presumably as a result of radical coupling.

In a variation of the alkylation procedure,[68,71] the reactions were induced with visible light using a combination of an α-diketone, such as biacetyl or camphorquinone, and a peroxide, such as di-*tert*-butyl peroxide. When used alone α-diketones failed to initiate the photoalkylation reaction, probably because the ability of excited state diketones to abstract hydrogen is rather weak. Presumably, when used in combination, the α-diketone absorbs visible light, then reacts with the peroxide to produce the hydrogen abstracting agent.

Obata and Niimura[72] used only di-*tert*-butyl peroxide with ultraviolet light as the photoinitiator. They reported the oxidative dimerization of methyl *N*-acetylglycinate **12a** and methyl pyroglutamate **12b** to give the corresponding dehydrodimers **14a** and **14b**, in yields of 51% and 64%, respectively. In a separate study,[21] analogous reactions of *N*-benzoylglycine methyl ester **12c** and *N*-benzoylalanine methyl ester **12d** afforded the corresponding dimers **14c**, in 37% yield, and **14d**, in 20% yield. In these reactions **12c** also afforded **15c** in 34% yield, and **12d** gave **15d** in 10% yield. The production of **14a–d, 15c**, and **15d** may be rationalized as shown in Scheme 2. Presumably methylation of the amino acid derivatives **12c** and **12d** results from coupling of the corresponding electrophilic α-carbon-centered radicals **13c** and **13d** with nucleophilic methyl radical, produced by β-scission of *tert*-butoxy radical.

Over the past decade there have been a number of reports on the reaction of glycine derivatives with *N*-bromosuccinimide (NBS) under free-radical reac-

Scheme 1

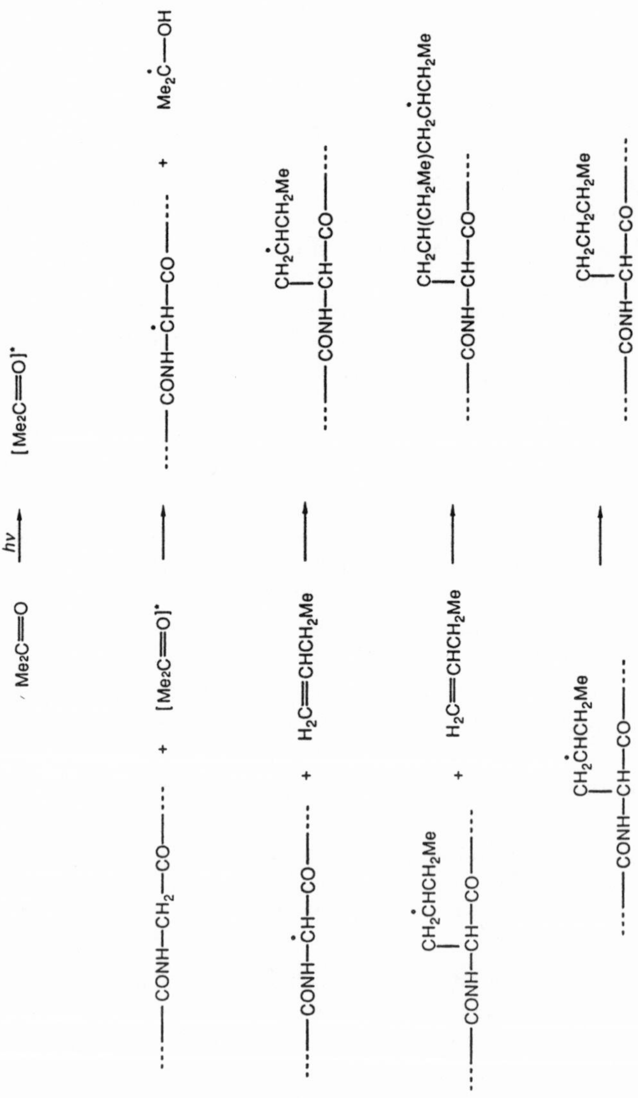

$Me_2C{=}O \xrightarrow{h\nu} [Me_2C{=}O]^{\bullet}$

$----CONH{-}CH_2{-}CO---- + [Me_2C{=}O]^{\bullet} \longrightarrow ----CONH{-}\overset{\bullet}{C}H{-}CO---- + Me_2\overset{\bullet}{C}{-}OH$

$----CONH{-}\overset{\bullet}{C}H{-}CO---- + H_2C{=}CHCH_2Me \longrightarrow$

$\begin{array}{c} \overset{\bullet}{C}H_2CHCH_2Me \\ | \\ ----CONH{-}CH{-}CO---- \end{array}$

$\begin{array}{c} \overset{\bullet}{C}H_2CHCH_2Me \\ | \\ ----CONH{-}CH{-}CO---- \end{array} + H_2C{=}CHCH_2Me \longrightarrow$

$\begin{array}{c} CH_2CH(CH_2Me)CH_2\overset{\bullet}{C}HCH_2Me \\ | \\ ----CONH{-}CH{-}CO---- \end{array}$

$\begin{array}{c} CH_2\overset{\bullet}{C}HCH_2Me \\ | \\ ----CONH{-}CH{-}CO---- \end{array} \longrightarrow$

$\begin{array}{c} CH_2CH_2CH_2Me \\ | \\ ----CONH{-}CH{-}CO---- \end{array}$

88

Scheme 1 (continued)

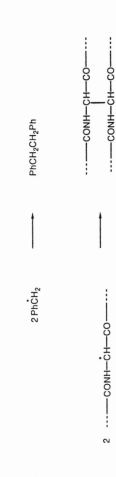

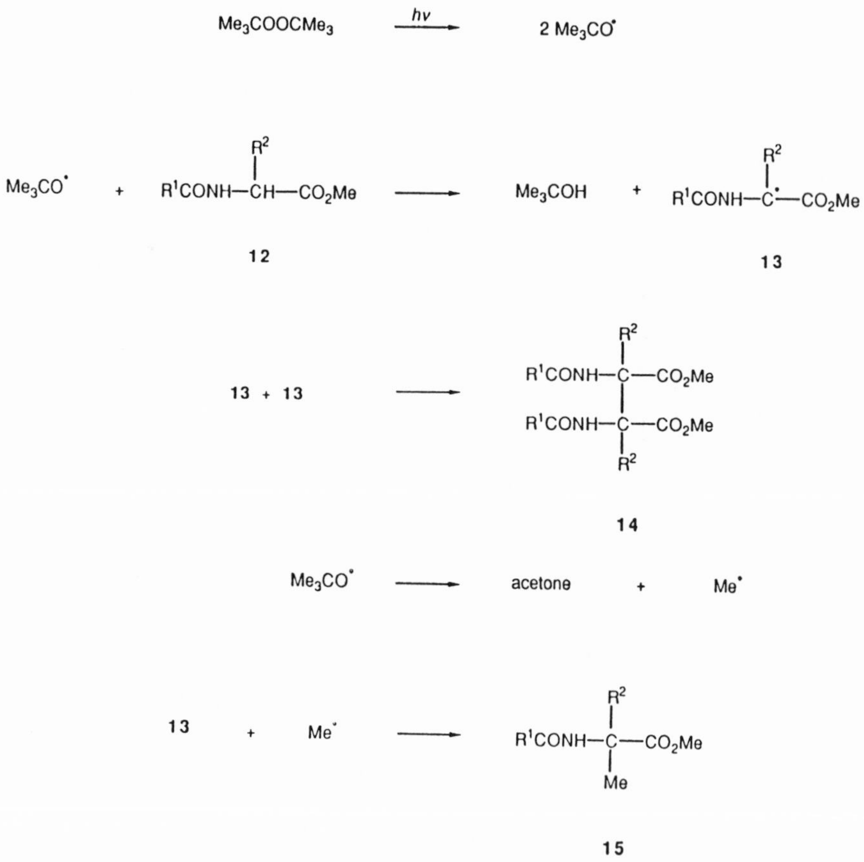

$$Me_3COOCMe_3 \xrightarrow{h\nu} 2\ Me_3CO^{\bullet}$$

$$Me_3CO^{\bullet} \quad + \quad \underset{\textbf{12}}{R^1CONH-\overset{\overset{\displaystyle R^2}{|}}{C}H-CO_2Me} \quad \longrightarrow \quad Me_3COH \quad + \quad \underset{\textbf{13}}{R^1CONH-\overset{\overset{\displaystyle R^2}{|}}{\underset{}{C}}{}^{\bullet}-CO_2Me}$$

$$\textbf{13 + 13} \quad \longrightarrow \quad \begin{array}{c} R^1CONH-\overset{\overset{\displaystyle R^2}{|}}{C}-CO_2Me \\ R^1CONH-\underset{\underset{\displaystyle R^2}{|}}{C}-CO_2Me \end{array}$$

$$\textbf{14}$$

$$Me_3CO^{\bullet} \quad \longrightarrow \quad acetone \quad + \quad Me^{\bullet}$$

$$\textbf{13} \quad + \quad Me^{\bullet} \quad \longrightarrow \quad R^1CONH-\overset{\overset{\displaystyle R^2}{|}}{\underset{\underset{\displaystyle Me}{|}}{C}}-CO_2Me$$

$$\textbf{15}$$

a) $R^1 = Me, R^2 = H$
b) $R^1, R^2 = CH_2CH_2$
c) $R^1 = Ph, R^2 = H$
d) $R^1 = Ph, R^2 = Me$

Scheme 2

$$12c \xrightarrow{\ Br^{\bullet}\ } 13c \xrightarrow{\ Br_2\ } 16a$$

Scheme 3

tion conditions.[73-80] For example, treatment of the glycine derivative **12c** with NBS in carbon tetrachloride, with irradiation to initiate the reaction, gave the α-bromoglycine derivative **16a**.[75] Since the reaction works equally well with

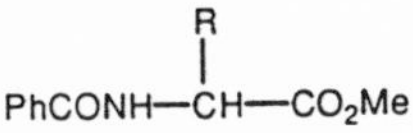

$$\overset{\displaystyle R}{\underset{\displaystyle}{\text{PhCONH}-\overset{|}{\text{CH}}-\text{CO}_2\text{Me}}}$$

16

a) R = Br
b) R = Cl
c) R = OCOPh

bromine in place of NBS,[74,75] the mechanism most probably involves a bromine atom chain (Scheme 3).

α-Carbon-centered radicals are likely intermediates in a variety of other free-radical substitution reactions of glycine derivatives. Treatment of the glycine derivative **12c** with sulfuryl chloride in carbon tetrachloride, with benzoyl peroxide to initiate the reaction, gave the α-chloroglycine derivative **16b**.[81] Reaction of **12c** with *tert*-butyl perbenzoate in benzene, in the presence of cupric octanoate, gave the benzoate **16c**.[82] Presumably, each of these reac-

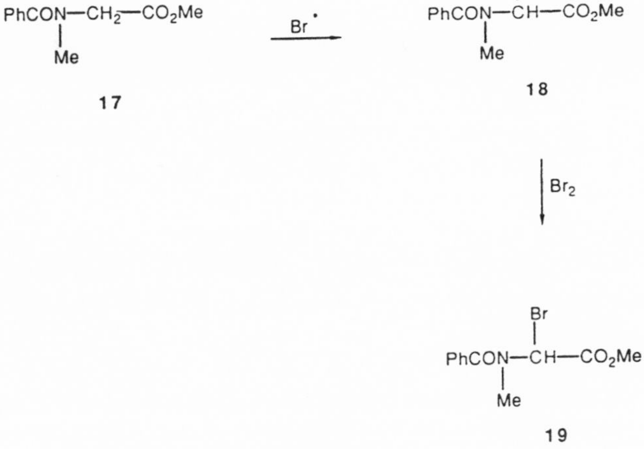

$$\underset{\displaystyle \text{Me}}{\overset{\displaystyle}{\text{PhCON}-\text{CH}_2-\text{CO}_2\text{Me}}} \qquad \xrightarrow{\ Br^{\bullet}\ } \qquad \underset{\displaystyle \text{Me}}{\overset{\displaystyle}{\text{PhCON}-\overset{\bullet}{\text{CH}}-\text{CO}_2\text{Me}}}$$

17   **18**

$$\Big\downarrow Br_2$$

$$\underset{\displaystyle \text{Me}}{\overset{\displaystyle Br}{\text{PhCON}-\overset{|}{\text{CH}}-\text{CO}_2\text{Me}}}$$

19

Scheme 4

tions involves the glycyl radical **13c** as an intermediate, generated by hydrogen atom transfer from **12c**. Recently, the free-radical carboxylation of glycine derivatives has also been reported.[83]

In reactions of NBS with derivatives of amino acids other than glycine, α-carbon-centered radicals are also implicated as intermediates. The sarcosine derivative **17** gave the bromide **19** on treatment with NBS, presumably via the α-carbon-centered radical **18** (Scheme 4).[84] In other cases the mechanism of reaction has been more difficult to elucidate. The valine derivative **20a** reacted with excess NBS to give the dibromide **21**.[84,85] From the relative rates of reaction of **20a** and the deuteriated analogs **20b** and **20c**, it was established that there is a deuterium isotope effect of 3.7 for cleavage of the α-carbon–hydrogen bond, but no deuterium isotope effect for reaction at the β position. On this basis the

20	21

a) $R^1 = R^2 = H$
b) $R^1 = D$, $R^2 = H$
c) $R^1 = H$, $R^2 = D$

mechanism proposed for the reaction is as shown in Scheme 5. Hydrogen atom transfer from **20a** affords the radical **22**, which reacts by bromine atom incorporation to give the bromide **23**. Subsequent elimination of hydrogen bromide from **23** gives the *N*-acylimine **24**, which undergoes tautomerism to give **25**. Addition of bromine to **25** affords the dibromide **21**. The bromine is produced through reaction of hydrogen bromide with NBS. In support of the proposed mechanism, both **24** and **25** were detected in crude reaction mixtures.

Again in the reactions of methyl pyroglutamate **12b** and *N*-benzoylalanine methyl ester **12d** with NBS, there is evidence that reaction occurs via the corresponding α-carbon-centered radicals **13b** and **13d**. In each case a deuterium isotope effect has been measured for α-carbon–hydrogen bond cleavage.[86]

A common feature of the hydrogen-atom transfer reactions of amino acid derivatives is the selectivity for reaction of glycine residues.[30,37,65–71,83] Elad et al.[65–71] observed the selective reaction of glycine residues in the photoalkylation of peptides. This general trend is reflected by the reactions of **26** and **27** with toluene, on irradiation with ultraviolet light in the presence of acetone.[69] The selectivity for alkylation of the glycine residue was 7 : 1 for **26** and 20 : 1 for **27**. With but-1-ene as the alkylating agent, the selectivity was 10 : 1 for **26** and

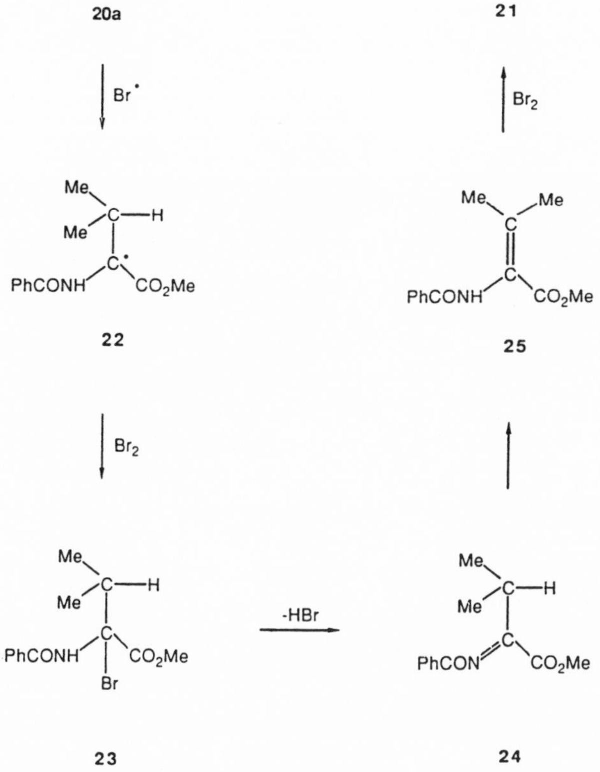

20a   21

Scheme 5

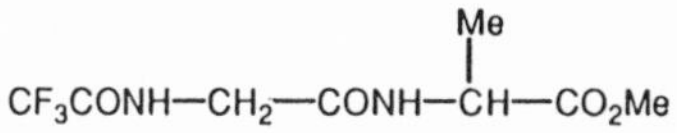

26 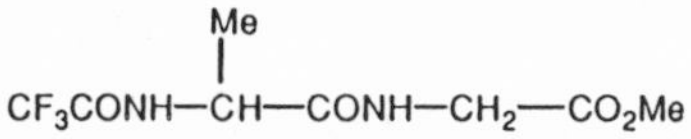   27

7 : 1 for **27**.[69] The selective photoalkylation of glycine residues in lysozyme[70] and other polypeptides[68–71] has also been reported.

Reactions of dipeptide derivatives with NBS are also selective for glycine residues.[19,85] For example, treatment of **28a** and **29a** with NBS in dichloromethane gave the corresponding bromides **28b** and **29b**. The bromides **28b** and **29b** were characterized by conversion to the corresponding methoxyglycine derivatives **28c** and **29c**, which were isolated in yields of 65% and 73%, based on **28a** and **29a**, respectively.

The preferential reactivity of glycine residues toward hydrogen atom abstrac-

Me   Me
        \   /
R   CH   Me   Me
| | \   /
PhCONH—CH—CONH—CH—CO₂Me   CH   R
                        · | |
                    a) R = H   PhCONH—CH—CONH—CH—CO₂Me
                    b) R = Br
            28   c) R = OMe   29

tion is contrary to the expectation that tertiary radicals should be formed in preference to secondary ones.[87] Glycine residues afford secondary radicals by α-carbon–hydrogen bond homolysis, whereas analogous reactions of derivatives of other amino acids, such as alanine and valine, produce tertiary radicals. This anomaly was investigated by studying reactions of mixtures of amino acid derivatives with NBS and di-*tert*-butyl peroxide.[85,86] With each reagent the rate of formation of the radical **13c** by hydrogen atom transfer from the glycine derivative **12c** was found to be faster than the rate of reaction of the alanine derivative **12d** to give **13d**, which was in turn faster than the rate of production of the radical **22** by hydrogen transfer from the valine derivative **20a**. To the extent that thermodynamic criteria control the pathways and rates of free-radical reactions, these results indicate that, in direct contrast to expectation, the secondary radical **13c** is marginally more stable than the tertiary radical **13d**, and both **13c** and **13d** are considerably more stable than **22**.

This peculiar stability of the radical **13c** can be attributed to a particularly favorable geometry. Stabilization of the radicals **13c**, **13d**, and **22** will result from overlap of the semioccupied p orbital with the π orbitals of the amido and methoxycarbonyl substituents. There will be maximum overlap of these orbitals in planar conformations of the radicals **13c**, **13d**, and **22**. The radical **13d** will be destabilized compared to **13c** by nonbonding interactions associated with planar conformations of **13d**, and **22** will be even less stable owing to more severe nonbonding interactions (Figure 3). These destabilizing influences outweigh the normal thermodynamic preference for the production of tertiary radicals.

The relative rates of production of the radicals **13d** and **18** in reactions with NBS and di-*tert*-butyl peroxide are very similar. This supports the hypothesis that the rate of hydrogen atom transfer from amino acid derivatives is affected by the extent of nonbonding interactions in the product radicals, since the degree of nonbonding interactions in planar conformations of **13d** and **18** is also very similar (Figure 3).

The rates of reaction of **12b** with NBS and di-*tert*-butyl peroxide are faster than the corresponding rates of reaction of the glycine derivative **12c**, consistent with the rationale outlined above. The radical **13b** can adopt planar conformations that are relatively free of nonbonding interactions (Figure 3). Formation of the radical **13b** is favored by the relief of ring strain and by the release of steric interactions between the methoxycarbonyl substituent and the β hydrogens in **12b**. It is possible that formation of the radical **13b** is also favored

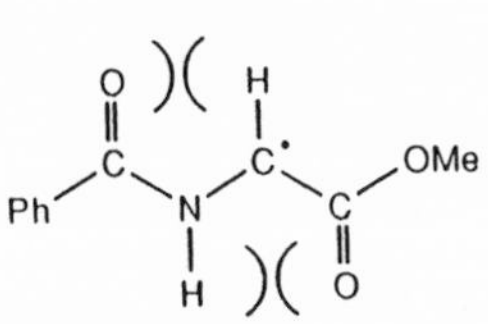

13c

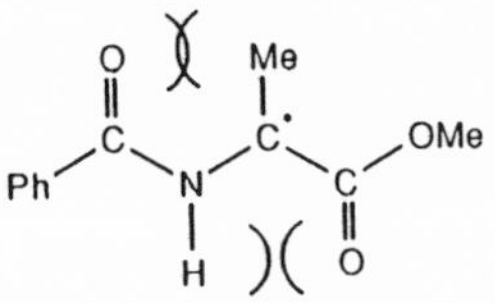

13d

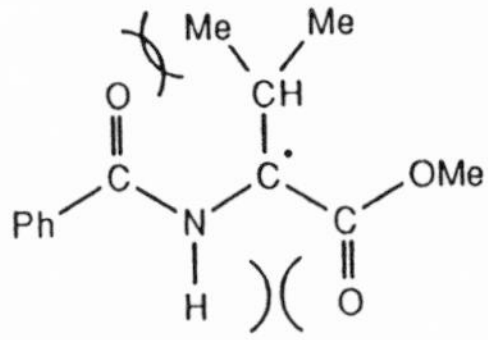

22

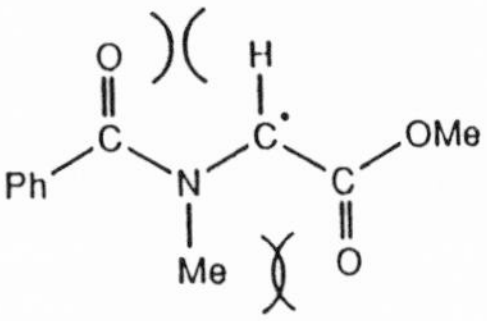

18

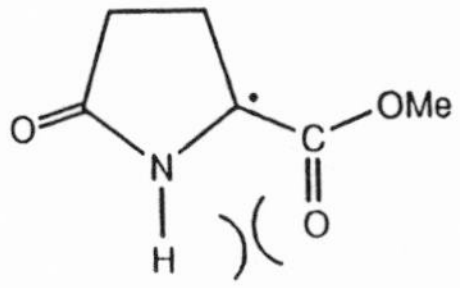

13b

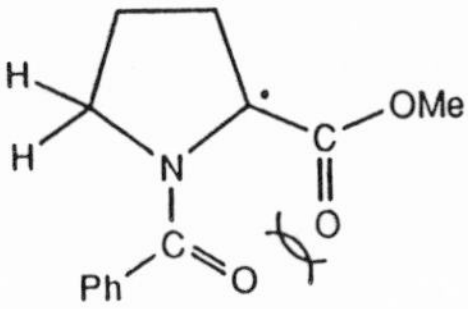

30

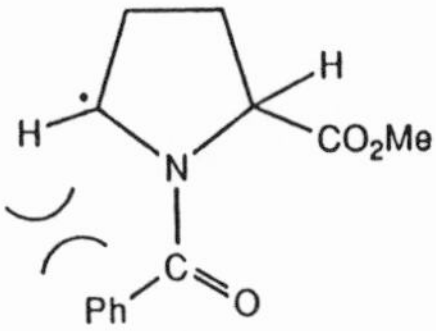

31

Figure 3. Nonbonding interactions in planar conformations of radicals derived from amino acid derivatives.

entropically by the inflexibility of the ring in **12b**, holding the amido group and the α carbon in the planar orientation as required for stabilization of the product radical **13b**. Thus the radical **13b** is more stable than the glycyl radical **13c** and this is reflected in the relative rates of reaction of **12b** and **12c** with bromine atom and *tert*-butoxy radical.

Based on this hypothesis, reaction of N-benzoylproline methyl ester **32** to give the α-carbon-centered radical **30** would be expected to be much slower than the rate of reaction of the glycine derivative **12c** to give **13c**, because the nonbonding interactions are much more severe in **30** than in **13c** (Figure 3). While the rate of reaction of **32** is faster than the rate of reaction of **12c**, the rate of formation of **30** is considerably slower than the rate of formation of **13c**. In

32   33

fact, steric interactions associated with planar conformations of the radical **30** are so severe that the predominant reaction of **32** is to produce the radical **31**, instead of **30**. Analogous regioselectivity has been observed in an electrochemical reaction of N-methoxycarbonylproline methyl ester **33**.[88] Presumably the fact that selective reactions of derivatives of pyroglutamic acid and proline have not been observed in biological systems is due to the relatively rare natural occurrence of these amino acids, compared to that of glycine.

In summary, the relative rates of reaction of **12b**, **12c**, **12d**, **17**, **20a**, and **32** with NBS and di-*tert*-butyl peroxide indicate that the selective reaction of glycine residues in these and other free-radical reactions of amino acid derivatives is due to the stability of the radicals produced by atom transfer reactions. Radicals formed by hydrogen abstraction from N-acylglycine derivatives may adopt planar conformations that are relatively free of nonbonding interactions and in which there is maximum delocalization of the unpaired electron. Radicals produced by similar reactions of derivatives of other amino acids such as alanine and valine are relatively unstable because of nonbonding interactions.

The bioactivation of many peptide hormones involves the oxidation of a C-terminal glycine-extended precursor to yield both an α-amidated peptide and glyoxylic acid (Scheme 6).[89,90] The reaction is catalyzed by the enzyme peptidylglycine α-amidating monooxygenase, which requires copper ions, L-ascorbate, and molecular oxygen. The reaction proceeds via an intermediate α-hydroxyglycine derivative,[91] presumably formed from the corresponding glycyl radical. It is interesting to speculate that the substrates of the

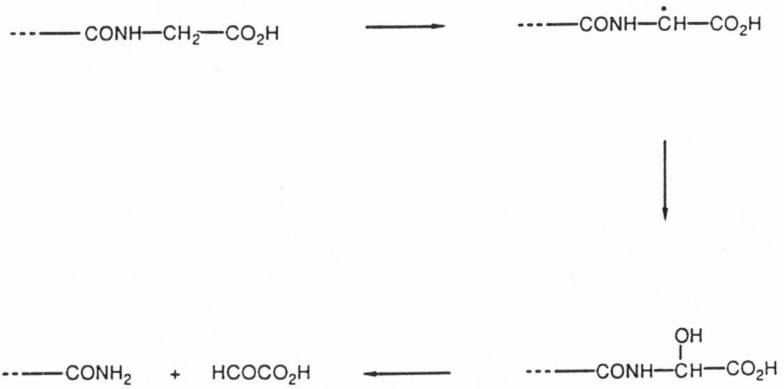

··· ——CONH—CH$_2$—CO$_2$H   ⟶   ··· ——CONH—ĊH—CO$_2$H

$$\downarrow$$

OH
|
··· ——CONH$_2$ +   HCOCO$_2$H   ⟵   ··· ——CONH—CH—CO$_2$H

Scheme 6

monooxygenase enzyme are synthesized with glycine at the C terminus because the glycine residue is so easily removed by oxidation.

The regioselectivity of hydrogen transfer reactions of amino acid derivatives is affected by other factors apart from the selectivity for reaction of glycine residues. For example, changes to the amino and carboxyl groups affect the reactivity of an amino acid derivative and the stability of the corresponding α-carbon-centered radical. One example of this phenomenon, mentioned previously, was the selective reaction of C-terminal amino acid residues in dipeptides, as studied by ESR spectroscopy.[37,40–42,62] The ammonium group strongly deactivates the adjacent α position of N-terminal amino acid residues toward hydrogen atom abstraction.

With protected dipeptides reaction has been observed to occur at both the C-terminal and N-terminal residues. The photoalkylation of **34** with toluene, using acetone and ultraviolet light as the photoinitiator, gave **35** and **36** in yields of 27% and 25%, respectively.[67] A similar reaction of **9**, mentioned previously, gave **10b**, in 26% yield, and **11b**, in 35% yield.[67] The reaction of **37a** with NBS in dichloromethane, with irradiation to initiate the reaction, gave only the bromide **37b**, characterized by conversion to the corresponding methoxyglycine derivative **37c**, which was isolated in 62% yield based on **37a**.[23] In direct contrast, the phthaloyl derivative **38a** gave only **38b**, from which **38c** was produced and isolated in 74% yield.[23] Neither **39** nor **40**, nor their derivatives, were detected in the reactions of **37a** and **38a**, respectively.

There is no definitive explanation for the regioselective reaction of the N-terminal amino acid residue of **37a**. It is possible that the selectivity reflects a preferred conformation adopted by **37a** in dichloromethane. It has been reported previously[92,93] that protected dipeptides adopt preferred conformations owing to hydrogen bonding in nonpolar solvents. These conformational effects

MeCONH—CH$_2$—CONH—CH$_2$—CO$_2$Me

34

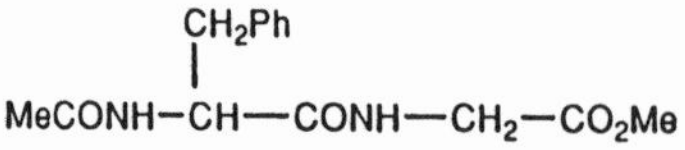

CH$_2$Ph
|
MeCONH—CH—CONH—CH$_2$—CO$_2$Me

35

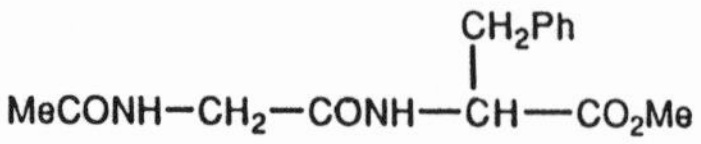

CH$_2$Ph
|
MeCONH—CH$_2$—CONH—CH—CO$_2$Me

36

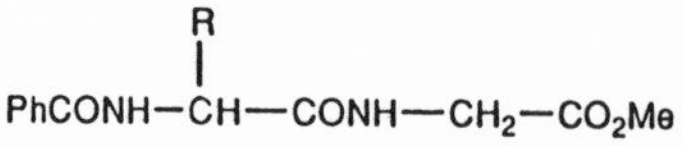

R
|
PhCONH—CH—CONH—CH$_2$—CO$_2$Me

37

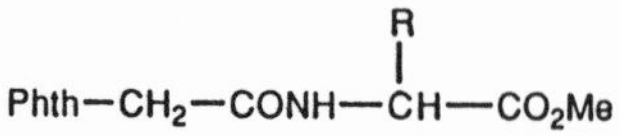

R
|
Phth—CH$_2$—CONH—CH—CO$_2$Me

38

a) R = H
b) R = Br
c) R = OMe

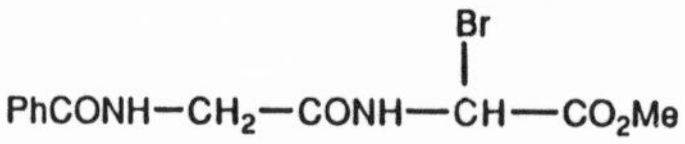

Br
|
PhCONH—CH$_2$—CONH—CH—CO$_2$Me

39

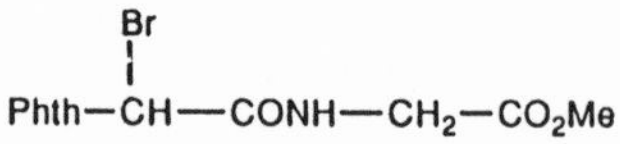

Br
|
Phth—CH—CONH—CH$_2$—CO$_2$Me

40

Phth =

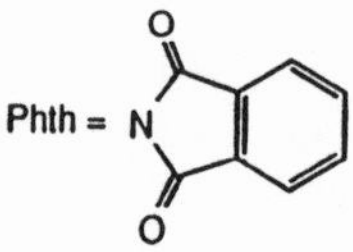

are absent in polar solvents, hence the relative lack of regioselectivity in the reactions of **9** and **34**, carried out in the presence of acetone.

The regioselective bromination of the C-terminal amino acid residue in **38a** can be attributed to the effect of the phthaloyl substituent. The α position of an *N*-phthaloyl-substituted amino acid derivative is less reactive than that of an *N*-acylamino acid derivative toward reaction with bromine atom. This may be attributed to the relative stability and ease of formation of the corresponding

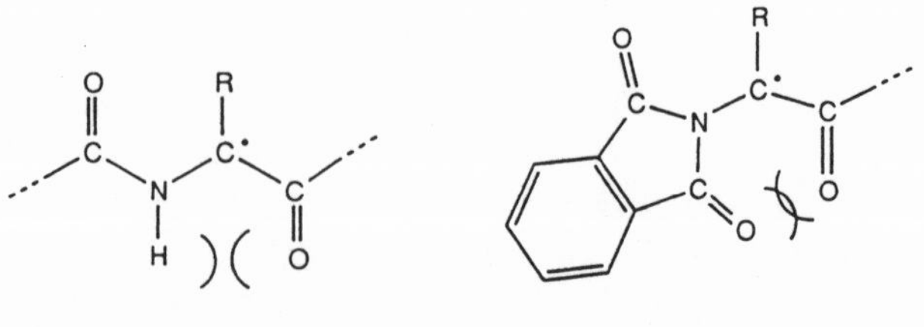

Figure 4. Nonbonding interactions associated with planar conformations of amido- and pthalimido-substituted radicals.

α-carbon-centered radicals. Whereas acylamino-substituted radicals are stabilized by resonance, there is less delocalization of unpaired spin density by a phthalimido substituent. This is the case particularly with α-carbon-centered radicals derived from amino acid derivatives. In planar conformations of the radicals, where there is maximum resonance stabilization, the nonbonding interactions are considerably greater with phthalimido-substituted radicals than with acylamino-substituted radicals (Figure 4). The phthaloyl substituent is also likely to hinder approach of bromine atom to the *N*-terminal amino acid residue in **38a**.

The combined effect of the phthaloyl substituent and the selectivity for reaction of glycine residues is illustrated by the selective formation of the

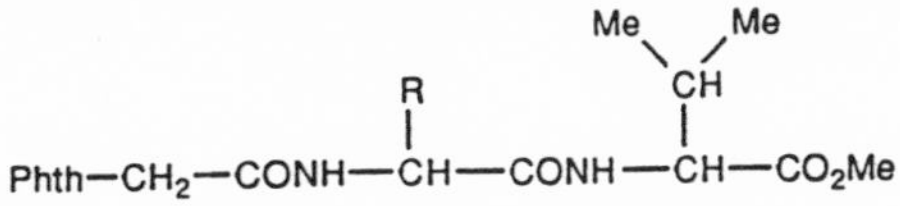

Me   Me
\\ /
R   CH
| |
Phth—CH_2—CONH—CH—CONH—CH—CO_2Me

41

a) R = H
b) R = Br
c) R = OMe

Me   Me
\\ /
CH
|
Phth—CH_2—CONH—ĊH—CONH—CH—CO_2Me

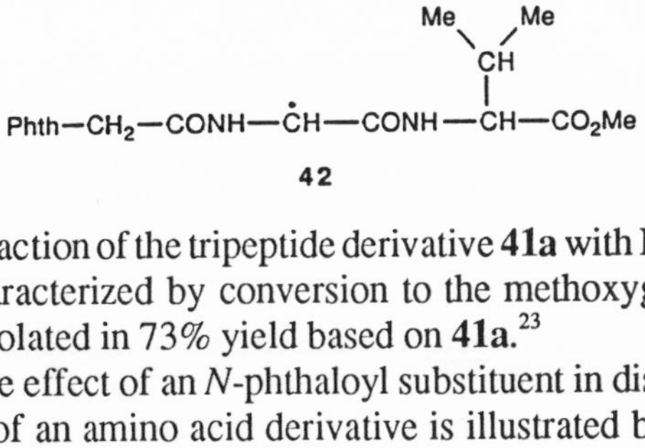

42

radical **42** in the reaction of the tripeptide derivative **41a** with NBS. The reaction afforded **41b**, characterized by conversion to the methoxyglycine derivative **41c**, which was isolated in 73% yield based on **41a**.[23]

The extent of the effect of an *N*-phthaloyl substituent in disfavoring reaction at the α position of an amino acid derivative is illustrated by the reactions of **43a** and **44a** with NBS. The brominated amino acid derivatives **43b** and **44b** were produced in yields of 87% and 83%, respectively, via the corresponding β-carbon-centered radicals **45** and **46**.[23] The regioselectivity observed in these reactions is contrary to that discussed above for reactions of *N*-acylamino acid derivatives with NBS. For example, the reaction of **20a** with bromine atom occurs at the α position to give the intermediate radical **22**.

Reactions of the valine derivative **20a** with sulfuryl chloride gave mixtures of the β-chlorovaline derivative **47** and the diastereomers of the γ-chlorovaline derivative **48**.[10,94] Presumably the peroxide-initiated chlorination of **20a** proceeds by initial hydrogen atom abstraction to give the radicals **49** and **50**, with subsequent chlorine incorporation at the sites of hydrogen abstraction. For

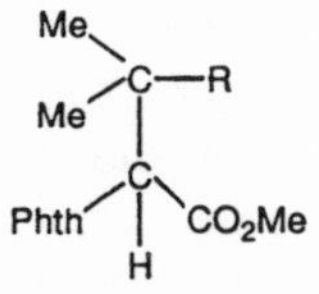

Me Me C—R Phth H CO₂Me

43

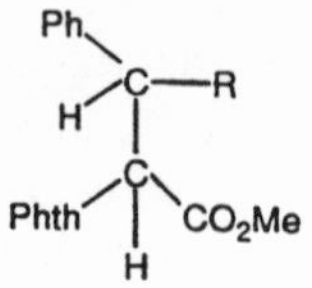

Ph H C—R Phth H CO₂Me

44

a) R = H
b) R = Br

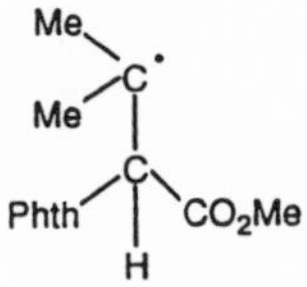

Me Me C• Phth H CO₂Me

45

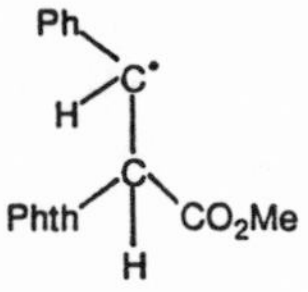

Ph H C• Phth H CO₂Me

46

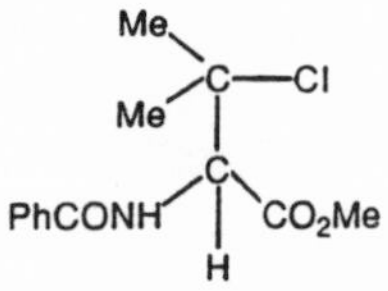

Me Me C—Cl PhCONH H CO₂Me

47

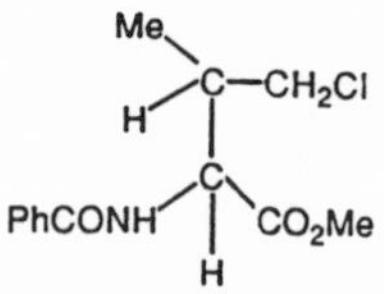

Me H C—CH₂Cl PhCONH H CO₂Me

48

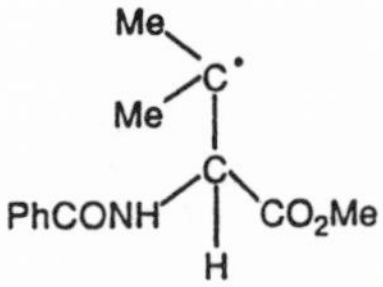

Me Me C• PhCONH H CO₂Me

49

Me H C—ĊH₂ PhCONH H CO₂Me

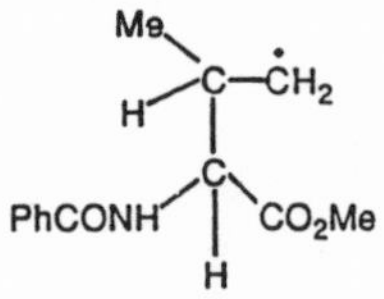

50

reaction in benzene the selectivity for β- to γ-carbon–hydrogen bond homolysis was 9 : 1.

The contrast between the reaction of **20a** with NBS via the α-carbon-centered radical **22** and that with sulfuryl chloride via the β- and γ-centered radicals **49** and **50** may be interpreted in terms of the relative degrees of carbon–hydrogen bond homolysis in the reaction transition states. With little development of radical character in the transition state of the chlorination reaction, the regioselectivity in this case is controlled by the inductive electron-withdrawing effect of the amido and carboxy substituents acting to retard attack at the adjacent α position by electrophilic radicals involved in the hydrogen atom abstraction, thus favoring reaction at the β and γ positions. The reaction with NBS is more sensitive to radical stability effects since there is a greater degree of development of radical character in the transition state. Hydrogen atom transfer from the α position is favored, therefore, because the product radical

22 is stabilized by the combined effect of the resonance electron-donating amido and electron-withdrawing carboxy groups.

Reaction of N-benzoylsarcosine methyl ester **17** with sulfuryl chloride gave the chloride **51**.[84] Again this is in contrast to the reaction of **17** with NBS, in which the product was the bromide **19**. This contrast in regioselectivity in the reactions of **17** with NBS and sulfuryl chloride can also be attributed to the degrees of bond homolysis in the transition states of the respective reactions.

PhCON—CH$_2$—CO$_2$Me   PhCON—CH$_2$—CO$_2$Me
         | |
     CH$_2$Cl   ·CH$_2$

                  51   **52**

The development of appreciable radical character in the transition state of the reaction of **17** with NBS results in reaction via the amidocarboxy-substituted radical **18**, whereas the relative lack of development of radical character in the transition state of the reaction of **17** with sulfuryl chloride is manifest in regioselectivity determined by polar effects and resulting in reaction via the radical **52**.

The valine derivative **20a** reacts with di-*tert*-butyl peroxide to give the radicals **22** and **49**, whereas the sarcosine derivative **17** reacts via the radicals **18** and **52**.[84] With this reagent a balance between polar effects and the stability of the product radicals results in the competing reactions. These studies indicate that amidocarboxy-substituted radicals such as **22** and **18** are considerably more stable than, for example, **49** and **52**, but hydrogen atom transfer reactions may afford the less stable products if electrophilic radicals are involved in the hydrogen atom abstraction, and if there is little development of radical character in the transition state of the reaction.

Polar effects can outweigh the selectivity for reaction of glycine derivatives. Using a mixture of **20a** and **12c**, the reaction of the valine derivative **20a** with sulfuryl chloride, to give the radicals **49** and **50**, was found to be three times faster than reaction of **12c** to give the chloride **16b** via the radical **13c**.[95]

As a result of investigations[1–4] into the biosynthesis of isopenicillin N **54** from Arnstein's tripeptide **53**, a free-radical pathway has been proposed for the formation of the carbon–sulfur bond.[2] It has been suggested that this involves the formation of a radical such as **56**, by hydrogen transfer from the corresponding valine derivative **55**, with subsequent substitution at sulfur to give **54** (Scheme 7). Evidence in support of this proposal has been obtained from reactions of the enzyme isopenicillin N synthetase with modified substrates. For example, δ-(L-α-aminoadipoyl)-L-cysteinyl-D-allylglycine **57** reacted to give three products, **60**, **61**, and **62**.[4] These products are consistent with reaction via the intermediate allylic radical **59**, formed by hydrogen

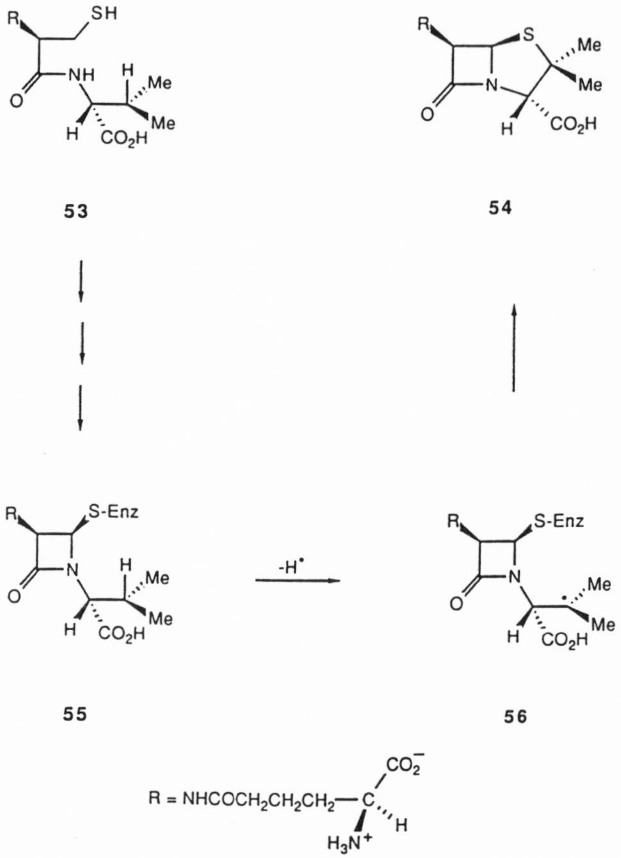

Scheme 7

transfer from **58** (Scheme 8). In a model system, treatment of the mercap-toazetidinone **63** with Udenfried's reagent afforded the penicillin derivative **64**.[11] By analogy with the hydrogen transfer reactions of **20a** discussed earlier, the regioselectivity of the reaction of **55** to give **56**, in preference to the α-carbon-centered radical **65**, can be attributed to polar effects. Similar effects may be involved in the regioselective β-hydroxylation of valine and N-methyl-valine residues in small peptides.[96–99]

In addition to polar and substituent effects, geometrical constraints on the transition state of the hydrogen transfer can affect the regioselectivity of reaction of amino acid derivatives. Photolysis of a solution of N-chloro-N-acetylnorvaline methyl ester **66a** in benzene afforded a mixture of the chlorides **67, 68**, and **69**, in the ratio circa 1.4 : 1 : 1.[100] These products are consistent with

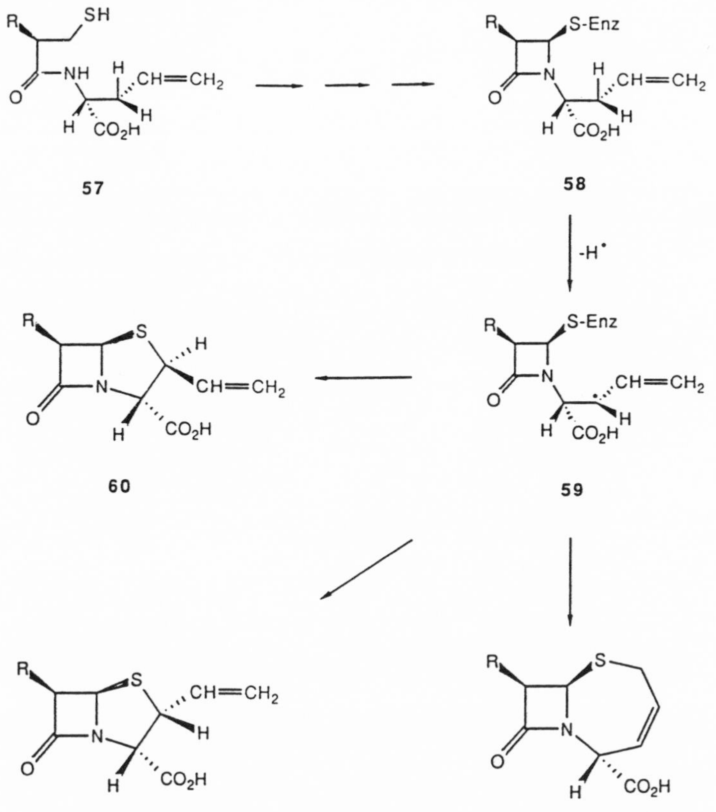

57 58 59 60 61 62

R = $NHCOCH_2CH_2CH_2$—$C(CO_2^-)(H)(^+NH_3)$

Scheme 8

intermolecular hydrogen transfer from **66a** or **66b** to chlorine atom. When 2,4,6-trimethylpyridine was used in the reaction, to react with hydrogen chloride and prevent the intermolecular process,[101] only the chloride **69** was produced. Under these conditions the production of **69** can be attributed to intramolecular 1,5-hydrogen transfer to the amido radical **70** (Scheme 9). The primary radical **71** is formed in preference to the secondary radicals **72** and **73** and the α-carbon-centered radical **74**, owing to the relative ease of formation of the six-membered ring transition state.[101–105] In another example, photolysis of the butyramide **75** in the presence of 2,4,6-trimethylpyridine gave the

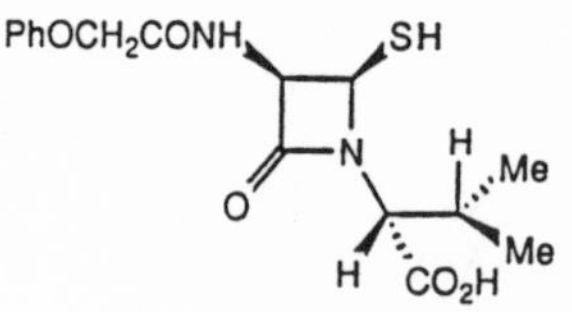

63 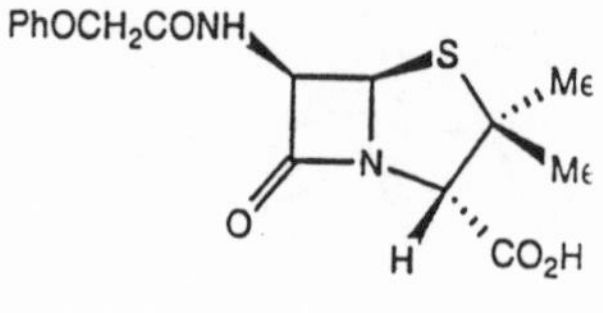   **64**

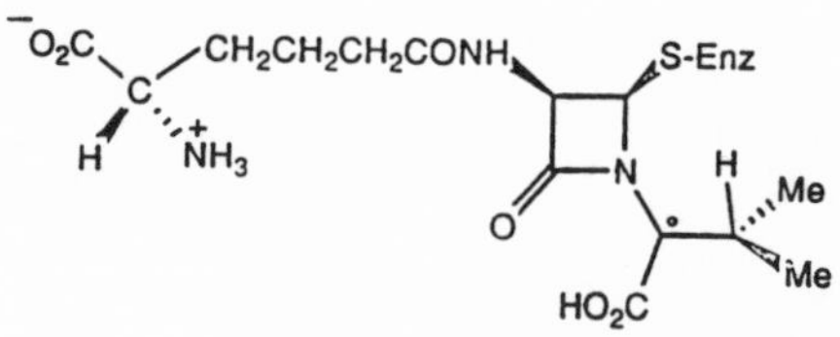

65

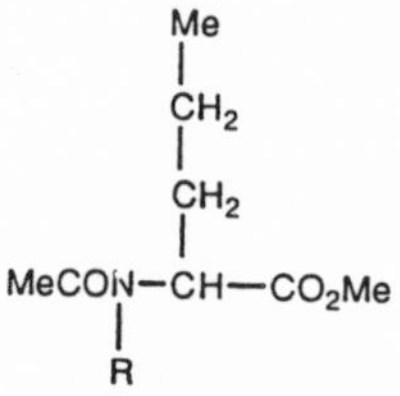

66
a) R = Cl
b) R = H

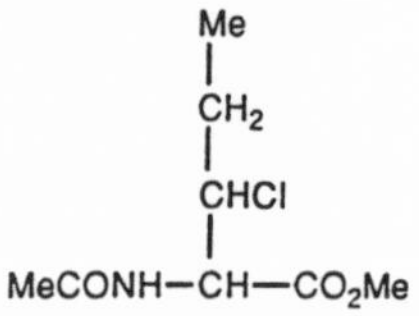

67

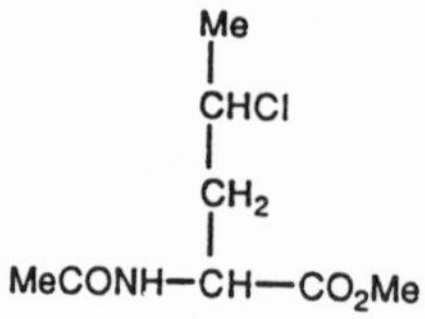

68 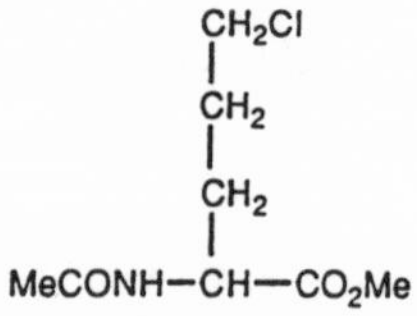   **69**

chloride **76**, consistent with regioselective intramolecular 1,5-hydrogen atom abstraction by the amido radical **77**, to give **78** (Scheme 10).[106]

Hydrogen transfer reactions of 2,5-diketopiperazines have attracted particular attention, in synthesis and as models in studies of reactions of proteins. Sperling and Elad[107] reported the photosensitized alkylation of diketopiperazines in reactions with alkenes. Hausler et al.[108] have studied the photochemically induced oxidation of cyclic dipeptides with molecular oxygen. Thus, the anhydrides of prolylproline **79a**, glycylproline **80a** and sarcosylproline **81a**, gave the corresponding hydroperoxides **79b**, **80b**, and **81b**, in yields of 50%, 32%, and 35%, respectively. The products **79b**, **80b**, and **81b** are consistent with an electron transfer mechanism, with reaction proceeding via the corresponding α-carbon-centered radicals **82**, **83**, and **84**. An interesting feature of the reactions of **80a** and **81a** is the selectivity for reaction

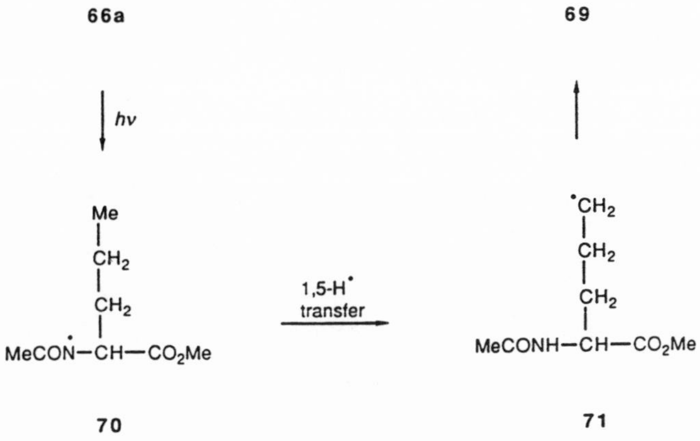

66a

69

$\downarrow$ *hv*

$\uparrow$

Me
|
CH$_2$
|
CH$_2$
|
MeCON$\overset{\bullet}{-}$CH$-$CO$_2$Me

$\xrightarrow{\text{1,5-H}^{\bullet}}$
transfer

$^{\bullet}$CH$_2$
|
CH$_2$
|
CH$_2$
|
MeCONH$-$CH$-$CO$_2$Me

70

71

Scheme 9

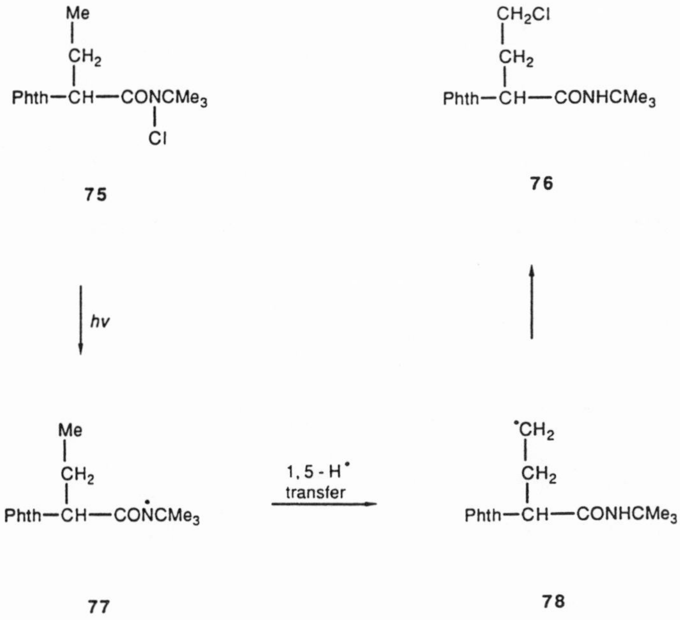

Me
|
CH$_2$
|
Phth$-$CH$-$CONCMe$_3$
|
Cl

CH$_2$Cl
|
CH$_2$
|
Phth$-$CH$-$CONHCMe$_3$

75

76

$\downarrow$ *hv*

$\uparrow$

Me
|
CH$_2$
|
Phth$-$CH$-$CO$\overset{\bullet}{N}$CMe$_3$

$\xrightarrow{\text{1,5 - H}^{\bullet}}$
transfer

$^{\bullet}$CH$_2$
|
CH$_2$
|
Phth$-$CH$-$CONHCMe$_3$

77

78

Scheme 10

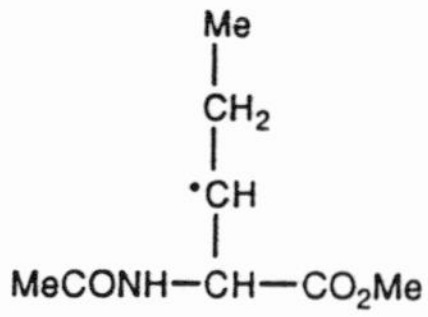

72

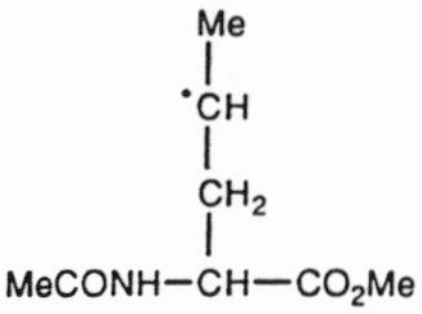

73

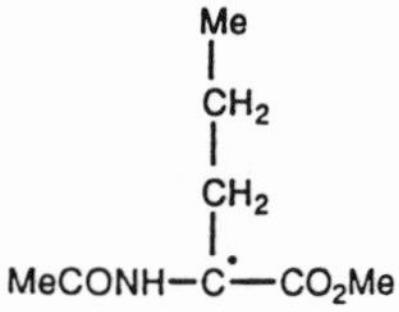

74

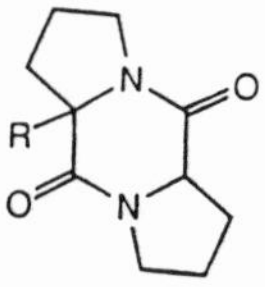

79

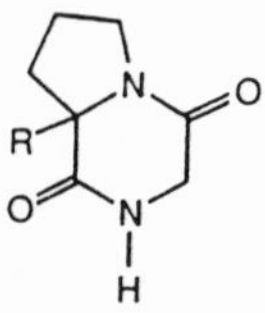

80

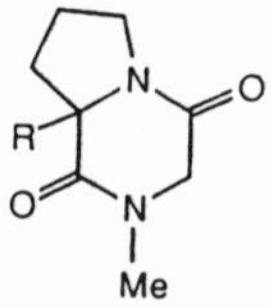

81

a) R = H
b) R = OOH

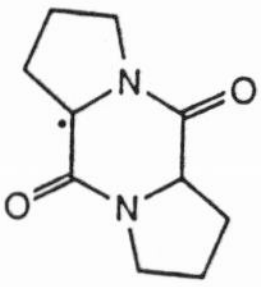

82

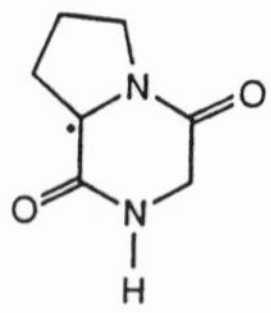

83

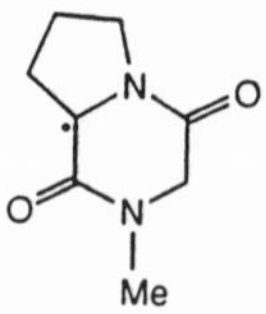

84

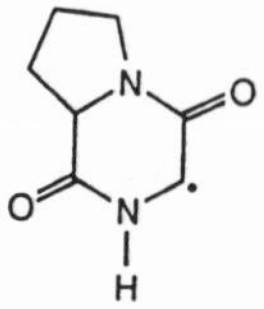

85

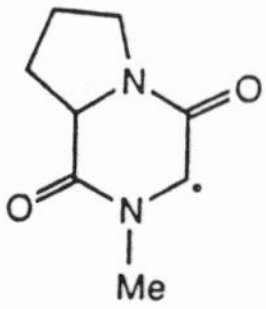

86

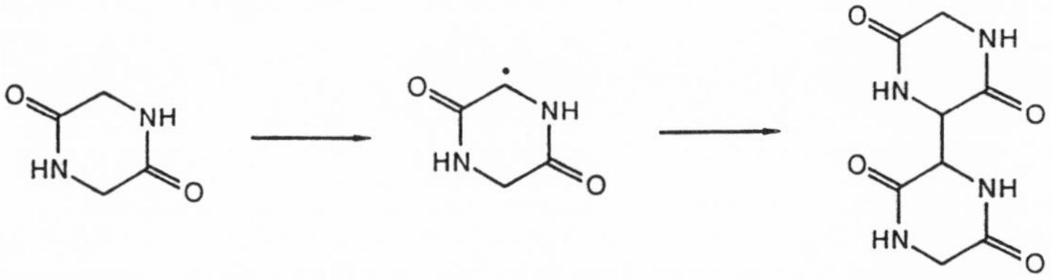

87   **88**   **89**

Scheme 11

at the α position of the proline residues. This is in contrast to the reactions of the proline derivative **32** with NBS and di-*tert*-butyl peroxide,[86] discussed previously, in which the radical **30** does not form. Whereas nonbonding interactions inhibit the formation of **30**, the radicals **83** and **84** can adopt planar conformations that are relatively free of such steric effects. Under these circumstances the tertiary amidocarboxy-substituted radicals **83** and **84** are formed in preference to the secondary radicals **85** and **86**.

In the absence of oxygen, irradiation of aqueous solutions of piperazine-2,5-dione **87** gave the dehydrodimer **89**, from coupling of the radical **88** (Scheme 11).[109] The dimer **89** was also formed in thermally and photochemically induced reactions of **87** with di-*tert*-butyl peroxide.[109]

There have been a number of reports of synthetic studies of the reactions of 2,5-diketopiperazines with bromine and NBS, under free-radical reaction conditions.[110–116] In one study,[115,116] irradiation with a 250-W mercury lamp of a mixture of sarcosine anhydride **90a** and NBS (2 equivalents), in dichloromethane at reflux under nitrogen, gave the dibromide **92a**. The dibromide **92a** was characterized by conversion to the dimethoxydiketopiperazine **92d**, which was isolated in 65% yield, based on **90a**. The diketopiperazines **90b** and **90c** gave the corresponding dibromides **92b** and **92c**, characterized by conversion to the derivatives **92e** and **92f**, which were isolated in yields of 60% and 55%, respectively. Treatment of **90a** with 0.85 molar equivalents of NBS gave a mixture of the monobromide **91a**, the dibromide **92a**, and unreacted starting material **90a**, in the ratio circa 15 : 1 : 5. Similar treatment of **90b** and **90c** gave mainly the corresponding monobromides **91b** and **91c**. The monobromides **91a–c** were characterized by conversion to the corresponding monomethoxydiketopiperazines **91d–f**, which were isolated in yields of 54%, 61%, and 41%, respectively.

A reasonable mechanism for the production of **91a–c** and **92a–c** is as shown in Scheme 12. Analysis of reaction mixtures showed that **90a–c** are approximately three times more reactive than the corresponding monobromides **91a–c** toward hydrogen atom transfer. The selective formation of the

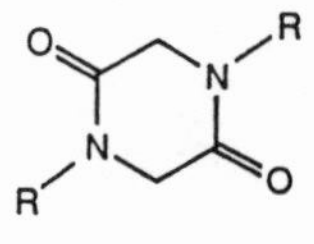

90

a) R = Me
b) R = CH₂Ph
c) R = CH₂Ph-*p*-OMe

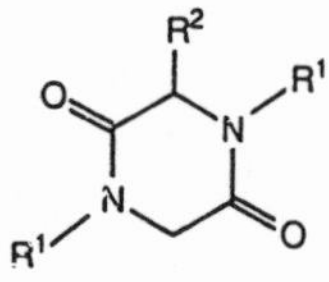

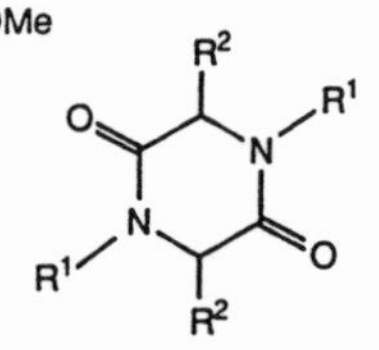

91

92

a) R^1 = Me, R^2 = Br
b) R^1 = CH₂Ph, R^2 = Br
c) R^1 = CH₂Ph-*p*-OMe, R^2 = Br
d) R^1 = Me, R^2 = OMe
e) R^1 = CH₂Ph, R^2 = OMe
f) R^1 = CH₂Ph-*p*-OMe, R^2 = OMe

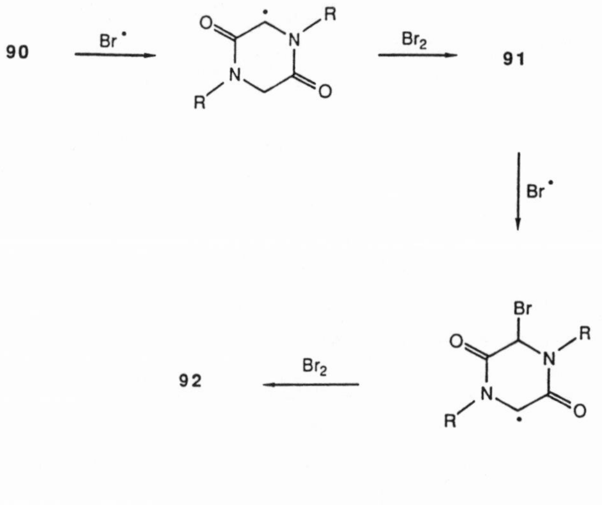

a) R = Me
b) R = CH₂Ph
c) R = CH₂Ph-*p*-OMe

Scheme 12

monobromides **91a–c** is contrary to the report by Williams and Kwast[114] that bromination of diketopiperazines such as **90c** gives exclusively 3,6-dibrominated products.

Bromination of the asymmetric diketopiperazine **93a** gave **93b**.[116] The regioselectivity observed in this reaction is consistent with the selective reaction of glycine residues in peptides,[19,85,86] as discussed previously. Substituent

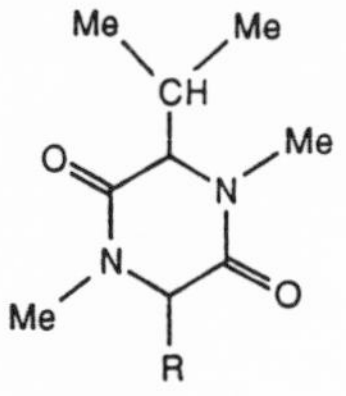

93

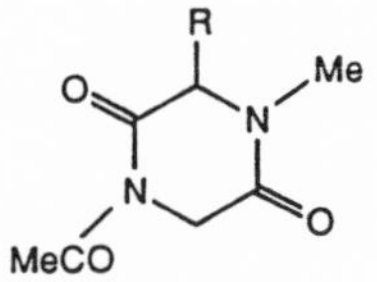

94

a) R = H
b) R = Br

effects have also been exploited in the regioselective bromination of diketopiperazines. The bromide **94b** was produced on treatment of **94a** with NBS.[116] By analogy with the effect of an N-phthaloyl substituent,[23] as discussed earlier, this may be attributed to the deactivating effect of the N-acetyl substituent in **94a**.

3. RADICALS FORMED IN REACTIONS WITH STANNANES

A number of studies of α-carbon-centered radicals have involved their generation by halogen transfer to stannyl radicals. Treatment of either of the α-haloglycine derivatives **16a** or **16b** with tributyltin hydride, in benzene or carbon tetrachloride, gave the reduced glycine derivative **12c** (Scheme 13).[81,85] Similarly, the α-bromosarcosine derivative **19** reacted to give **17**.[117] Presumably the selective halogen transfer from **16a**, **16b**, and **19** in the reactions carried out in carbon tetrachloride reflects the stability of the intermediate radicals **13c** and **18**. Less reactive substrates require solvents such as benzene that are inert to reaction with stannyl radicals. Evidence in support of the reaction mechanism was obtained by ESR spectroscopy. Irradiation with a 1000-W high-pressure mercury lamp of mixtures of di-*tert*-butyl peroxide, hexabutylditin, and the

$$ \textbf{16a,b} \xrightarrow{\text{Bu}_3\text{Sn}^\bullet} \textbf{13c} \xrightarrow{\text{Bu}_3\text{SnH}} \textbf{12c} $$

Scheme 13

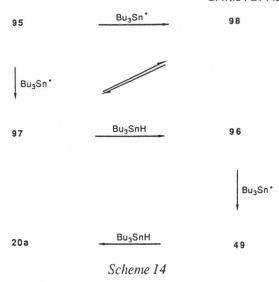

Scheme 14

bromides **16a** and **19** in the cavity of an ESR spectrometer gave doublet signals with further complex hyperfine splitting consistent with formation of the corresponding radicals **13c** and **18**.[117] Under these conditions photolysis of the peroxide produces *tert*-butoxy radical, which reacts with hexabutylditin to give tributylstannyl radical. The stannyl radical reacts with **16a** and **19** by bromine atom abstraction to give the radicals **13c** and **18**, respectively.

The α,β-dihalogenated amino acid derivatives **95a** and **95b** reacted with tributyltin hydride to give the corresponding β-halides **96a** and **96b**.[84,85,118] The production of **96a** and **96b** may be rationalized as shown in Scheme 14. Halogen atom transfer from the vicinal dihalides **95a** and **95b** would be expected to give the more stable of the possible corresponding product radicals **97a** and **97b**, and **98a** and **98b**. In the unlikely event that halogen atom abstraction gave as the first formed the less stable of the possible product radicals **97a** and **97b**, and **98a** and **98b**, a facile 1,2-halogen migration to give the thermodynamically more stable radical would be expected.[119] On this basis the formation of the β-halides **96a** and **96b** indicates that the amidocarboxy-substituted radicals **97a** and **97b** are more stable than the corresponding β-carbon-centered radicals **98a** and **98b**, as expected. Furthermore, the formation of only trace amounts of **20a**, the product of subsequent reduction of **96a** and **96b**, in the reactions with tributyltin hydride indicates that the radicals **97a** and **97b** are more stable than **49**.

It is interesting to note that while the typical reaction of vicinal dihalides with tributyltin hydride is to give the corresponding alkenes (Scheme 15),[120] there was no evidence of formation of the dehydro amino acid derivative **25** in the

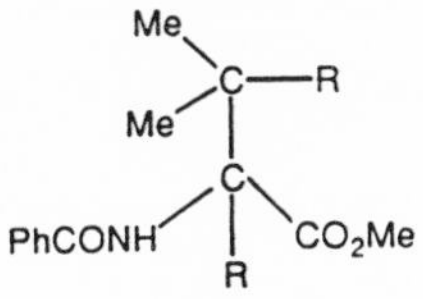

95

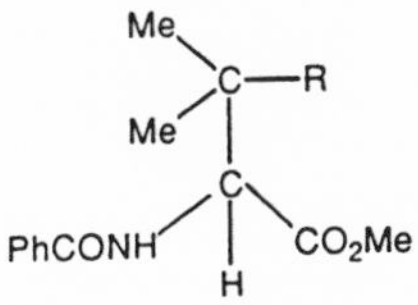

96

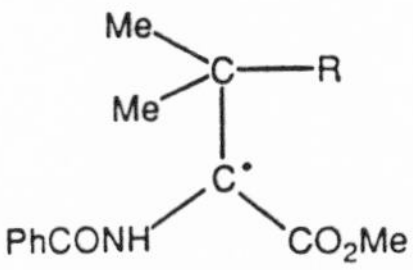

97

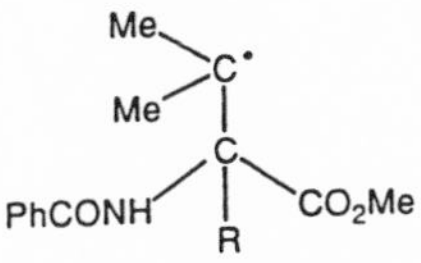

98

a) R = Br
b) R = Cl

reactions of **95a** and **95b**. Presumably the radicals **97a** and **97b** react by hydrogen atom abstraction to give **96a** and **96b**, respectively, rather than by β-scission to give **25**, owing to the destabilizing effect of steric interactions associated with **25** (Figure 5).

In variations of the reactions of halogenated amino acid derivatives with stannanes, tributyltin deuteride has been used to prepare regiospecifically deuteriated peptides[19,23,85] and other amino acid derivatives.[121] With hexabutyl-ditin the bromoglycine derivative **16a** reacted under irradiation to give a 1 : 1 mixture of the diastereomers of the dimer **14c**, in 67% yield.[21] Formation of **14c** may be attributed to coupling of the glycyl radical **13c** (Scheme 16). Irradiation of a mixture of **16a**, hexabutylditin, and dibenzyldisulfide afforded the amidothioketal **99** as the major product, presumably as a result of **13c** reacting with dibenzyldisulfide by substitution at sulfur (Scheme 16).[82]

There have been two independent reports of the synthesis of the allylglycine derivative **100** by homolytic allylation of the bromide **16a**. Treatment of **16a** with allyltributylstannane gave **100** in 62% yield,[19] while the reaction of **16a**

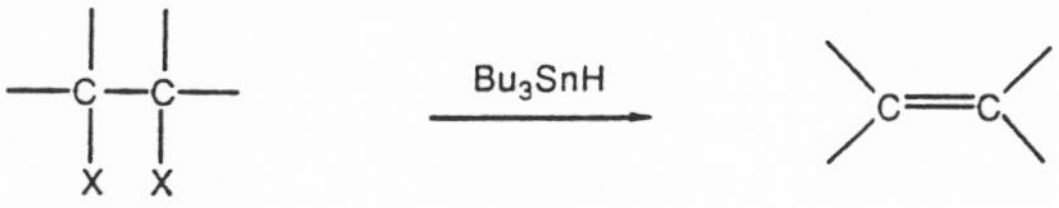

X = Cl, Br

Scheme 15

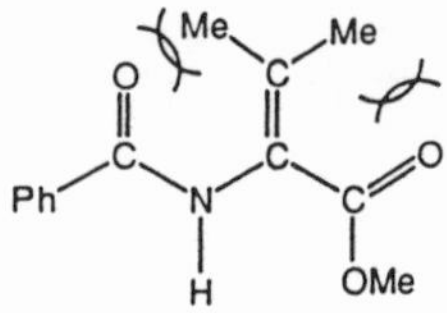

25

Figure 5.   Steric interactions in *N*-benzoyl-α,β-dehydrovaline methyl ester.

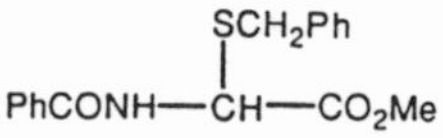

99 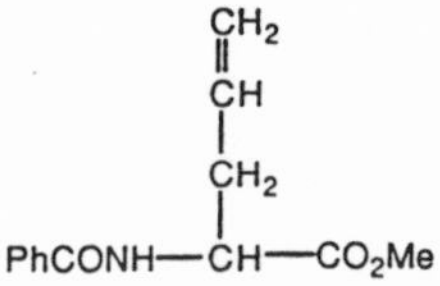   **100**

with allyltriphenylstannane gave **100** in 65% yield.[20] A probable mechanism
for these reactions is as shown in Scheme 17.

Baldwin et al.[20] reported reactions of the 2-substituted allyltributylstannanes
101a–c to give the corresponding 4-substituted allylglycine derivatives **102a–c**.
Allyl transfer reactions of **16a** with 1-, 2-, and 3-alkyl-substituted allyltributylstannanes have also been observed.[24] Treatment of the bromide **16a**
with the allylstannanes **103a–d** gave the corresponding glycine derivatives

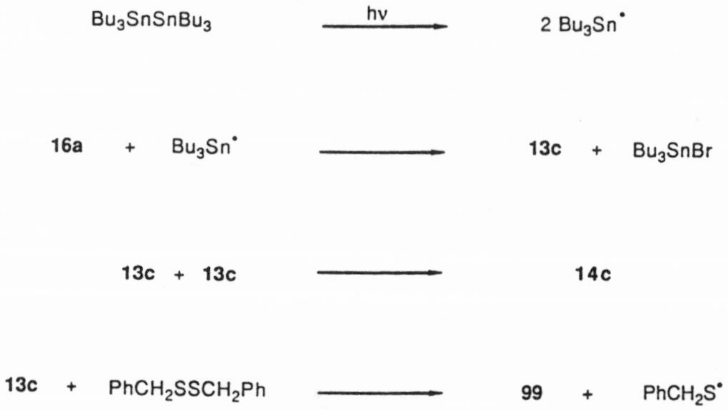

Scheme 16

16a   +   R$_3$Sn$^\bullet$ —————→   13c   +   R$_3$SnBr

13c   +   R$_3$SnCH$_2$CH=CH$_2$ —————→   100   +   R$_3$Sn$^\bullet$

R = Bu or Ph

Scheme 17

104a–d. There was no evidence of formation of the reduced glycine derivative **12c** in any of these reactions. The reactions to give **104a, 104c**, and **104d** are contrary to reports that hydrogen abstraction from 1- and 3-alkyl-substituted allylstannanes occurs in preference to allyl group transfer.[122,123] The radical **13c** reacts by addition to the stannanes **103a, 103c**, and **103d**, rather than by hydrogen atom abstraction (Scheme 18). Prior to this work, only Pereyre and

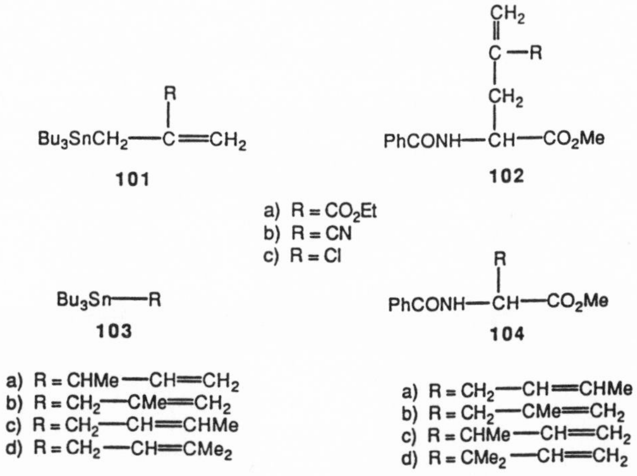

$$CH_2$$
$$\|$$
$$C-R$$
$$|$$
$$CH_2$$
$$|$$

Bu$_3$SnCH$_2$—C=CH$_2$ PhCONH—CH—CO$_2$Me

with R above the middle carbon

101   **102**

a) R = CO$_2$Et
b) R = CN
c) R = Cl

Bu$_3$Sn——R   PhCONH—CH—CO$_2$Me (with R above)

103   **104**

a) R = CHMe—CH=CH$_2$ a) R = CH$_2$—CH=CHMe
b) R = CH$_2$—CMe=CH$_2$ b) R = CH$_2$—CMe=CH$_2$
c) R = CH$_2$—CH=CHMe   c) R = CHMe—CH=CH$_2$
d) R = CH$_2$—CH=CMe$_2$ d) R = CMe$_2$—CH=CH$_2$

co-workers[124,125] had reported homolytic allyl transfer reactions of tributyl-(3-methylallyl)stannane **103c**. They noted that the allylation reaction is susceptible to polar effects, being faster with substrates which react via intermediate radicals substituted with electron-withdrawing groups. On this basis it seems likely that the allylation reaction is favored with electrophilic radicals such as **13c** and those used by Pereyre and co-workers, whereas nonpolar radicals react by hydrogen atom transfer.

The α-carbon-centered radical **13c** is also implicated as an intermediate in reactions of the benzoate **105** with stannanes. Treatment of **105** with tributyltin

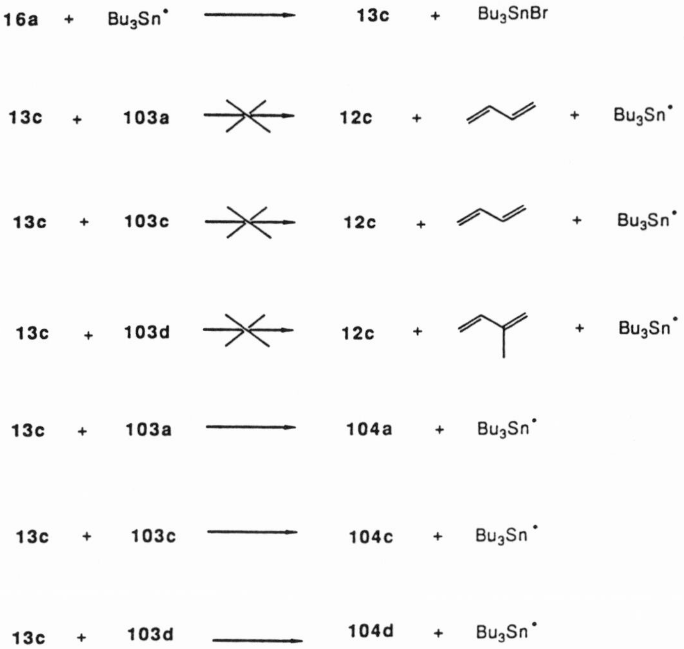

$$16a \; + \; Bu_3Sn^{\bullet} \longrightarrow 13c \; + \; Bu_3SnBr$$

$$13c \; + \; 103a \; \xmapsto{} \; 12c \; + \; \text{(propene)} \; + \; Bu_3Sn^{\bullet}$$

$$13c \; + \; 103c \; \xmapsto{} \; 12c \; + \; \text{(butene)} \; + \; Bu_3Sn^{\bullet}$$

$$13c \; + \; 103d \; \xmapsto{} \; 12c \; + \; \text{(isobutene)} \; + \; Bu_3Sn^{\bullet}$$

$$13c \; + \; 103a \longrightarrow 104a \; + \; Bu_3Sn^{\bullet}$$

$$13c \; + \; 103c \longrightarrow 104c \; + \; Bu_3Sn^{\bullet}$$

$$13c \; + \; 103d \longrightarrow 104d \; + \; Bu_3Sn^{\bullet}$$

Scheme 18

$$105 \; + \; Bu_3Sn^{\bullet} \longrightarrow 13c \; + \; Bu_3SnOCOPh$$

$$13c \; + \; Bu_3SnH \longrightarrow 12c \; + \; Bu_3Sn^{\bullet}$$

Scheme 19

$$Bu_3SnSnBu_3 \xrightarrow{h\nu} 2 \; Bu_3Sn^{\bullet}$$

$$105 \; + \; Bu_3Sn^{\bullet} \longrightarrow 13c \; + \; Bu_3SnOCOPh$$

$$13c \; + \; 13c \longrightarrow 14c$$

Scheme 20

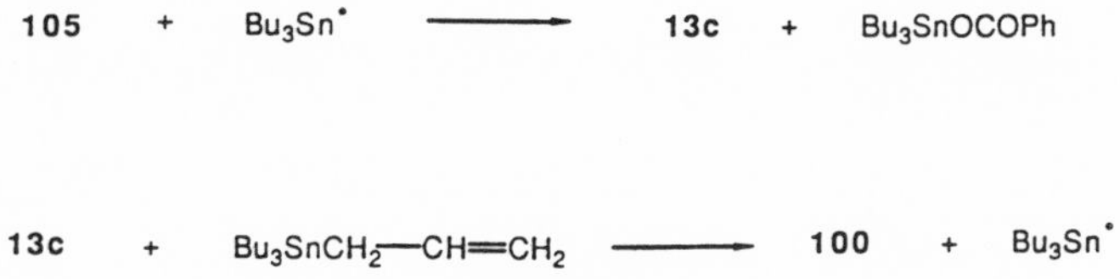

$$105 \quad + \quad Bu_3Sn^{\bullet} \quad \longrightarrow \quad 13c \quad + \quad Bu_3SnOCOPh$$

$$13c \quad + \quad Bu_3SnCH_2{-}CH{=}CH_2 \quad \longrightarrow \quad 100 \quad + \quad Bu_3Sn^{\bullet}$$

Scheme 21

$$106 \quad + \quad Bu_3Sn^{\bullet} \quad \longrightarrow \quad 13c \quad + \quad Bu_3SnOOCMe_3$$

$$13c \quad + \quad Bu_3SnH \quad \longrightarrow \quad 12c \quad + \quad Bu_3Sn^{\bullet}$$

Scheme 22

hydride gave the reduced glycine derivative **12c**, in 88% yield.[82] It seems likely that the mechanism of this reaction involves substitution of tributylstannyl radical on the benzoate **105** to give **13c** (Scheme 19). Although it is not a generally used method for the reduction of esters, there has been one other report of the reaction of benzoates with tributyltin hydride.[126] From that work it appears that the efficiency of the reaction depends on the stability of the

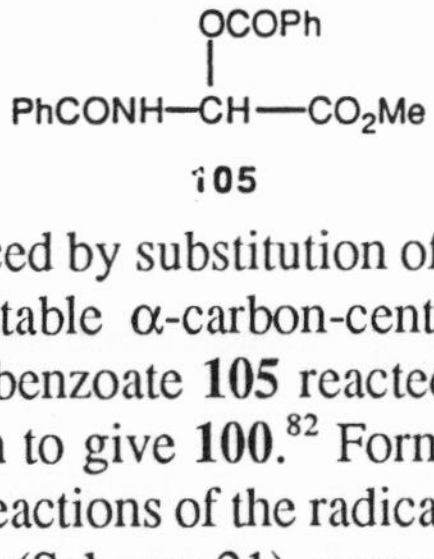

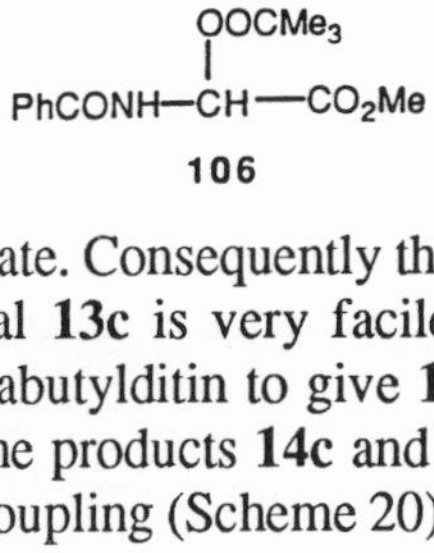

$$\begin{array}{cc} \overset{\displaystyle OCOPh}{\underset{|}{}} & \overset{\displaystyle OOCMe_3}{\underset{|}{}} \\ PhCONH{-}CH{-}CO_2Me & PhCONH{-}CH{-}CO_2Me \\ \mathbf{105} & \mathbf{106} \end{array}$$

radical produced by substitution of the benzoate. Consequently the reaction of **105** via the stable α-carbon-centered radical **13c** is very facile. In related reactions the benzoate **105** reacted with hexabutylditin to give **14c** and with allyltributyltin to give **100**.[82] Formation of the products **14c** and **100** may be attributed to reactions of the radical **13c**, by coupling (Scheme 20) and by allyl group transfer (Scheme 21), respectively.

The peroxide **106** reacted with tributyltin hydride to give **12c**.[82] A reasonable mechanism for this reaction is as shown in Scheme 22. The unusual mode of substitution of the peroxide by carbon–oxygen bond cleavage may be attributed to the stability of the product radical **13c**.

4. RADICALS FORMED BY ADDITION TO DEHYDRO AMINO ACID DERIVATIVES

Despite the considerable success that has been achieved with free-radical addition reactions of alkenes,[127] little attention has been given to the addition

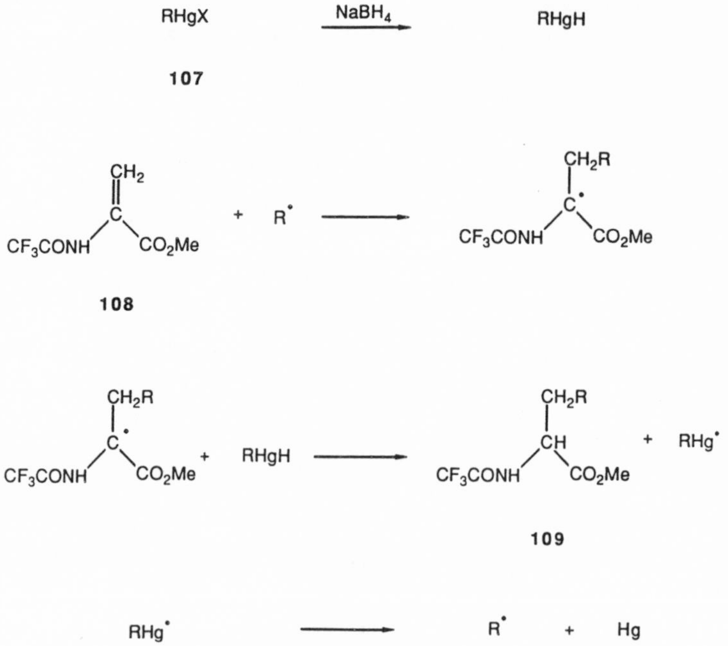

$$RHgX \xrightarrow{\quad NaBH_4 \quad} RHgH$$

107

CH₂=C(NHCOCF₃)CO₂Me + R· ⟶ CF₃CONH–C·(CH₂R)CO₂Me

108

CF₃CONH–C·(CH₂R)CO₂Me + RHgH ⟶ CF₃CONH–CH(CH₂R)CO₂Me + RHg·

109

$$RHg\cdot \longrightarrow R\cdot + Hg$$

a) R = CH₂CHMe₂, X = Br
b) R = c-C₆H₁₁, X = Cl
c) R = CHEtMe, X = Br
d) R = CMe₃, X = Cl
e) R = c-C₃H₅, X = Cl

Scheme 23

reactions of dehydro amino acid derivatives. The reactions of *N*-trifluoroacetyl-dehydroalanine methyl ester **108** with the alkylmercury halides **107a–e** and sodium borohydride to give **109a–e**, respectively, have been reported.[25] The mechanism proposed for these reactions is as shown in Scheme 23. Recently this work has been extended to reactions of dehydroalanine residues in di- and tripeptides.[27] The vitamin-B₁₂-photoelectrocatalyzed addition of alkyl bromides and carboxylic acids to methyl 2-acetamidoacrylate has also been reported.[26] Crossley and Reid[128] observed the reaction of *N*-phthaloyldehydroalanine methyl ester **111** with radicals generated by the action of tributyltin hydride on alkyl bromides and iodides. For example, heating the halides **110a** and **110b** with mixtures of **111** and tributyltin hydride in benzene at reflux with azobisisobutyronitrile to initiate reaction gave the corresponding adducts **112a** and **112b**, in yields of 79% and 74%, respectively (Scheme 24). These reactions

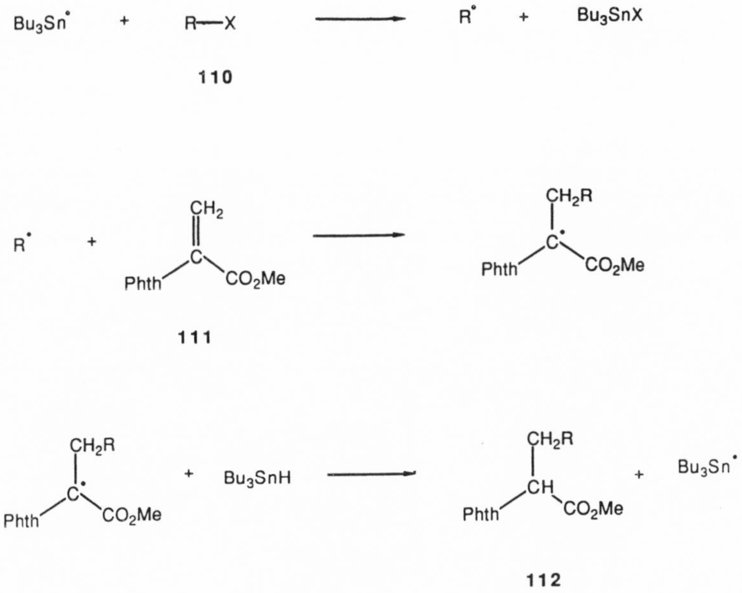

$Bu_3Sn^\bullet$ + R—X $\longrightarrow$ $R^\bullet$ + Bu_3SnX

110

$R^\bullet$ + [structure with CH₂, C, Phth, CO₂Me] → [structure CH₂R, C, Phth, CO₂Me]

111

[structure CH₂R, C, Phth, CO₂Me] + Bu_3SnH → [structure CH₂R, CH, Phth, CO₂Me] + $Bu_3Sn^\bullet$

112

a) R = CHMe₂, X = I
b) R = CH₂Ph, X = Br

Scheme 24

show that primary, secondary, and tertiary alkyl radicals add to α,β-dehydro amino acid derivatives to give the corresponding α-carbon-centered radicals. The direction of addition of radicals to dehydro amino acid derivatives can be attributed to the stability of the product α-carbon-centered radicals and to steric effects, with addition occurring at the less hindered end of the double bond.

5. RADICALS FORMED IN REACTIONS OF IMINES AND N-ACYLIMINES

The intramolecular addition of radicals to imines has been observed in the reactions of **113a** and **113b** with tributyltin hydride to give the corresponding rearranged products **116a** and **116b** (Scheme 25).[129,130] Presumably addition occurs in the *exo* mode to give **115a** and **115b**, in preference to the *endo* mode to give the corresponding α-carbon-centered radicals **114a** and **114b**, due to geometrical constraints on the transition state of the reaction. The reactions of **113a** and **113b** were studied as models of the biochemical interconversion of L-glutamic acid and L-threo-β-methylaspartic acid.[131] In this regard it is interesting to note that **113b** reacted with vitamin B₁₂ to give **116b**.[130]

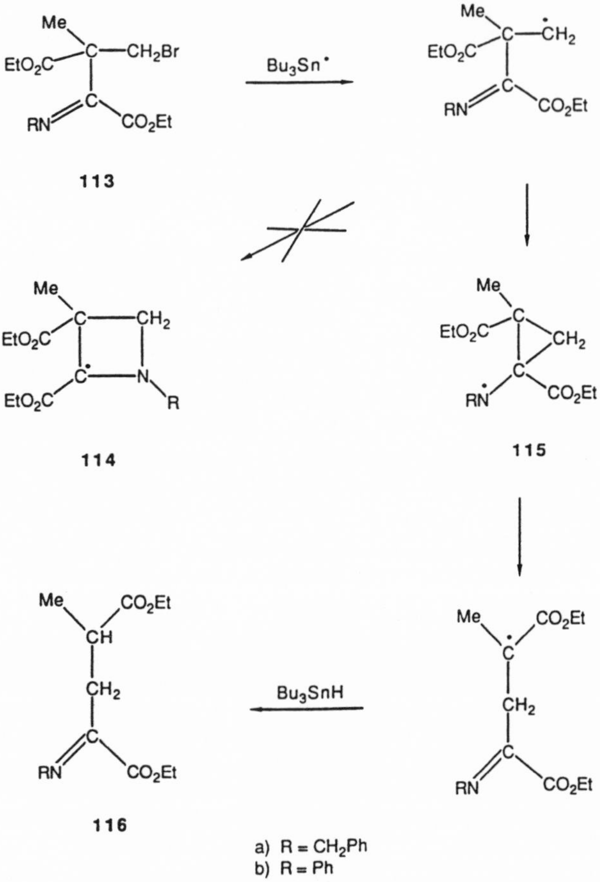

Scheme 25

Electron transfer reactions of *N*-acylimines are thought to be involved in the reactions of **16a** with the deprotonated nitroalkanes **117a–d** (2 equivalents).[22] High yields of the corresponding C-alkylated products **118a–d** were obtained in these reactions. In contrast, alkylation of alkyl nitronates generally takes place overwhelmingly on oxygen.[132–134] Other examples of C-alkylation of alkyl nitronates have been attributed to an electron transfer mechanism.[135,136] Accordingly, it is likely that **16a** reacts with the first mole equivalent of an alkyl nitronate such as **117a** to give the *N*-acylimine **119**. A subsequent electron transfer between **119** and the second mole equivalent of the alkyl nitronate **117a** gives a pair of free radicals that then combine (Scheme 26). It seems probable that the reactions of Grignard reagents with α-haloglycine derivatives[137,138] proceed via an analogous electron transfer mechanism.

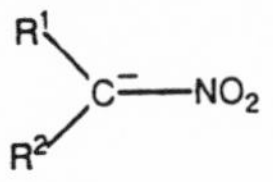

$$R^1 \diagdown C \overset{=}{\underset{R^2 \diagup}{}} NO_2$$

117

$$\begin{array}{c} R^1 \diagdown \\ R^2 \diagup \end{array} C-NO_2 \\ | \\ \underset{PhCONH \diagup \quad \diagdown CO_2Me}{CH}$$

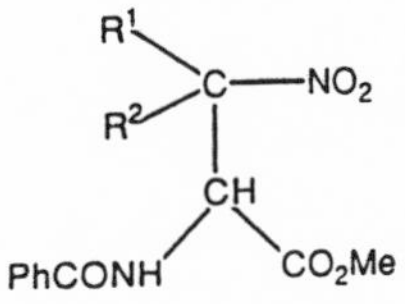

118

a) $R^1 = R^2 = H$
b) $R^1 = R^2 = Me$
c) $R^1 = Me, R^2 = H$
d) $R^1 = Ph, R^2 = H$

Photochemical reactions of the iminolactone **120a** and the iminolactam **120b** have also been studied.[139–142] Irradiation with a 450-W mercury lamp of **120a** and **120b** in 2-propanol gave the diastereomers of the dimers **122a** and **122b**, respectively, presumably via the corresponding α-carbon-centered radicals **121a** and **121b**. Each of the dimers **122a** and **122b** has an unusually weak carbon–carbon single bond and exists in solution at room temperature in equilibrium with the corresponding radical **121a** or **121b**. The radicals **121a** and **121b** have been identified by ESR spectroscopy and in spin trapping experiments.

The enthalpy of dissociation of the dimers **122a** and **122b** is solvent dependent.[139,142] For **122a** the enthalpy of dissociation is 22 kcal/mol in chloroform and 11 kcal/mol in ethanol.[139] The solvent effect has been rationalized in terms of the greater solvation in polar solvents of the polar radicals **121a** and **121b**

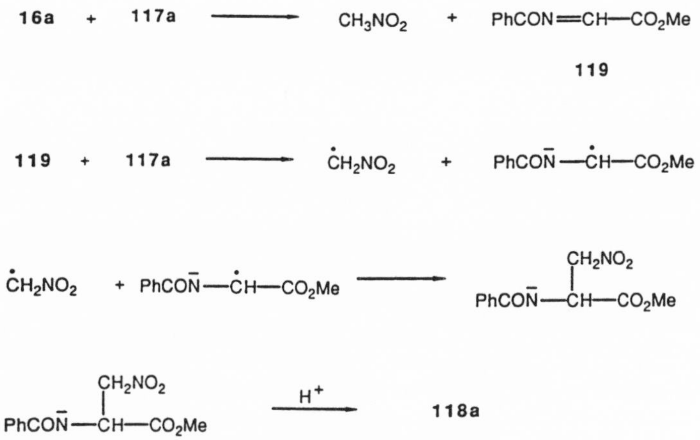

16a + **117a** $\longrightarrow$ CH_3NO_2 + $PhCON{=}CH{-}CO_2Me$

119

119 + **117a** $\longrightarrow$ $\overset{\bullet}{C}H_2NO_2$ + $PhCO\bar{N}{-}\overset{\bullet}{C}H{-}CO_2Me$

$\overset{\bullet}{C}H_2NO_2$ + $PhCO\bar{N}{-}\overset{\bullet}{C}H{-}CO_2Me$ $\longrightarrow$ $\underset{PhCO\bar{N}{-}CH{-}CO_2Me}{\overset{CH_2NO_2}{\overset{|}{}}}$

$\underset{PhCO\bar{N}{-}CH{-}CO_2Me}{\overset{CH_2NO_2}{\overset{|}{}}}$ $\overset{H^+}{\longrightarrow}$ **118a**

Scheme 26

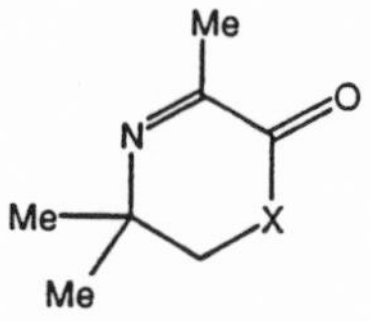

120

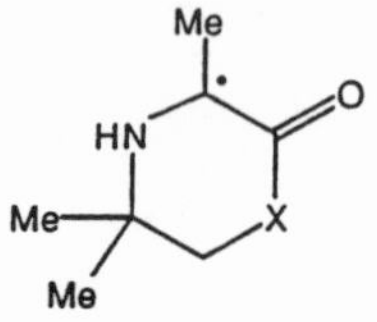

121

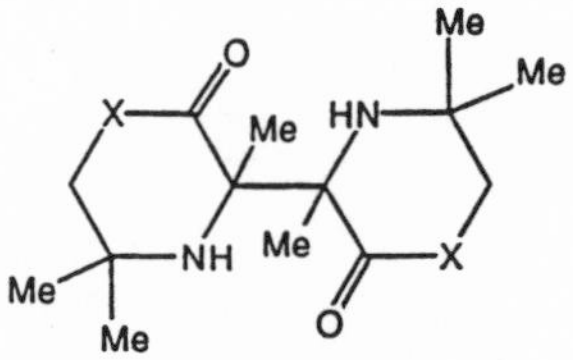

122

a) X = O
b) X = NH

compared to the corresponding dimers **122a** and **122b**. The dissociation of **122a** occurs more easily than that of **122b**. This has been attributed to the greater stability of the radical **121a** compared to **121b**, because the carboxyamido substituent in **121b** is less effective than the carboxyl group in **121a** as a resonance electron acceptor.[141]

The radicals **121a** and **121b** have been utilized in a number of studies as one-electron reducing agents.[141,143–149] In these processes the radicals **121a** and **121b** are oxidized to the iminolactone **120a** and the iminolactam **120b**, respectively. The reactions are thought to involve electron transfer from the radicals **121a** and **121b**, followed by proton transfer.

6. REACTIONS WITH ASYMMETRIC INDUCTION

There have been a number of reports of asymmetric induction in reactions of α-carbon-centered radicals derived from amino acid derivatives. Williams et al.[77] reported that the reduction of (–)-5(S),6(R)-**123** with tributyltin deuteride, followed by hydrogenolysis, gave (R)-**124** in 60% enantiomeric excess. Presumably the intermediate radical **125** adopts a conformation in which the phenyl substituents block one side to a lesser extent than the carbobenzoxy group blocks the opposite side. In this conformation tributyltin deuteride delivers the deuterium *cis* to the phenyl substituents. It should be noted that the asymmetric induction observed in the preparation of the bromide **123**[76] and other halogenated glycine derivatives[75,78,80,138] does not necessarily reflect the stereoselectivity of transfer of bromine atom to the corresponding intermediate radicals. It is more likely that the ratio of products reflects the relative stability

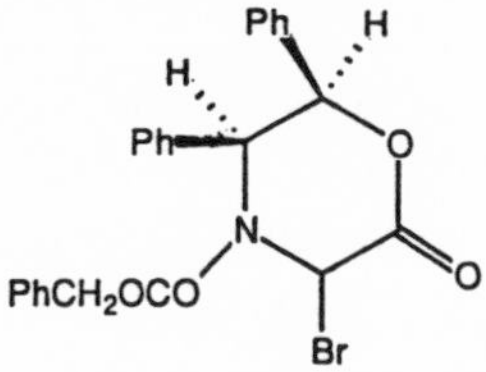

123

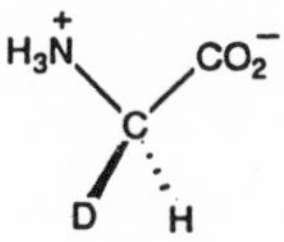

124

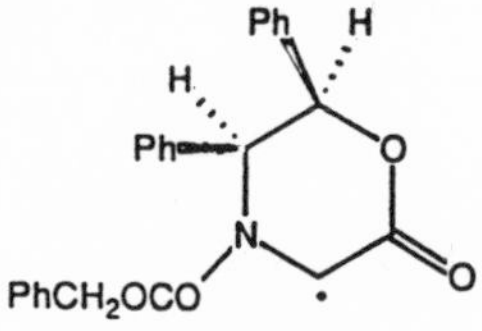

125

of the diastereomers because equilibration of α-halogenated glycine derivatives occurs readily.

The conversion of a glycine residue into a residue of an α-substituted amino acid generates a new chiral center. In the photoalkylation of peptides this occurs with a low degree of asymmetric induction.[66,67,69,71] For example, the

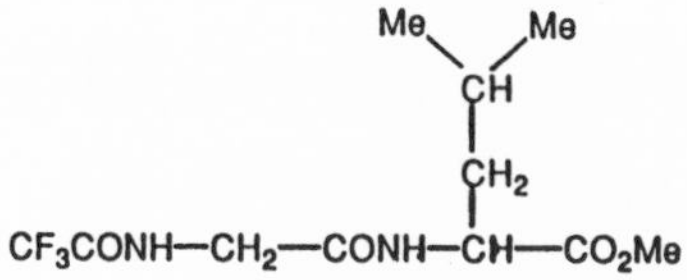

126

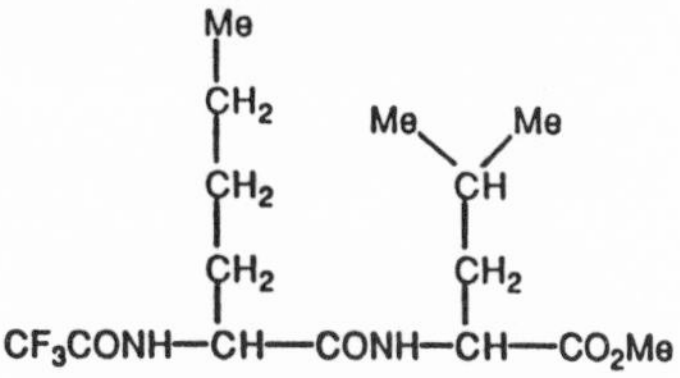

127

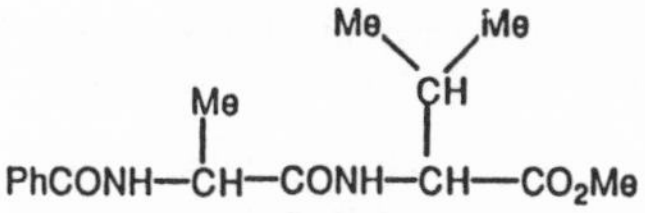

128

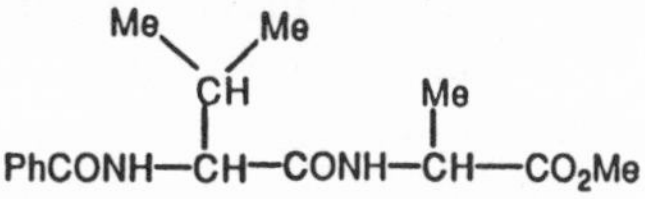

129

photoalkylation of (S)-N-trifluoroacetylglycylleucine methyl ester **126** with but-1-ene gave the (S),(S) and (S),(R) diastereomers of **127** in the ratio circa 6 : 4.[69] A similar degree of asymmetric induction was observed in the reactions of **28a** and **29a** with di-*tert*-butyl peroxide to give the diastereomers of **128** and **129**, respectively.[95]

A more significant degree of asymmetric induction has been observed in reactions of the brominated dipeptide **29b**, prepared from (R,S)-**29a**. Treatment of **29b** with allyltributyltin gave the (R,S),(S,R) and (R,R),(S,S) diastereomers of **130a** in the ratio circa 3 : 1.[19] The mixture of diastereomers of **130a** was characterized by hydrogenation to give **130b**, which compared with an authentic sample. Reaction of **29a** with the anion of 2-nitropropane gave the $(R,S),(S,R)$ and $(R,R),(S,S)$ diastereomers of **130c** in the ratio circa 3 : 1, characterized by reduction with tributyltin hydride to give **130d**.[95] Although there is no definitive explanation for the asymmetric induction observed in

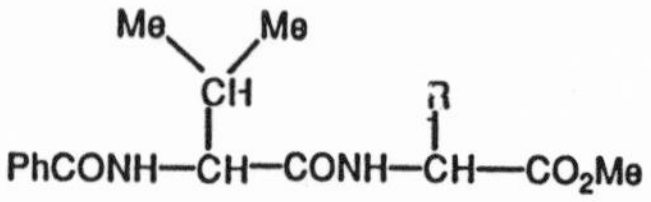

$$\underset{\text{Me}}{\overset{\text{Me}}{\diagdown}}\underset{\mid}{\overset{\mid}{\text{CH}}}$$

PhCONH—CH—CONH—CH—CO$_2$Me

130

a) R = CH$_2$CH=CH$_2$
b) R = CH$_2$CH$_2$Me
c) R = C(NO$_2$)Me$_2$
d) R = CHMe$_2$

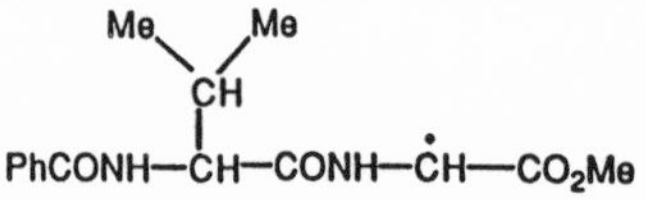

Me   Me
   \ /
   CH

PhCONH—CH—CONH—ĊH—CO$_2$Me

131

Me   Me
   \ /
   CH

PhCONH—CH—CON̄—ĊH—CO$_2$Me

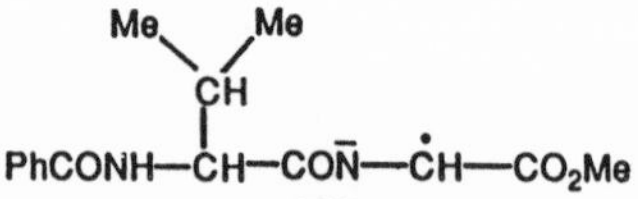

132

these reactions, it is possible that the stereoselectivity reflects preferred conformations of the radical **131** and the radical anion **132**, intermediates in the reactions of **29a** with allyltributyltin and deprotonated 2-nitropropane, respectively. Obviously the induction observed in these reactions warrants further investigation.

7. CONCLUSION

The foregoing survey illustrates factors that affect the formation and reaction of α-carbon-centered radicals derived from amino acids and their derivatives. The reactions include atom transfer, electron transfer, radical coupling, and radical addition processes. The reactions are affected markedly by the particular stability of the α-carbon-centered radicals, but substituent effects, polar effects, and geometrical constraints on reaction transition states also affect the reactions of these species. Each of these factors must be considered in order to understand the reactions of α-carbon-centered radicals. Although these radicals require further investigation in order that we can understand their reactions more completely, the results already obtained indicate the types of reactions that can be expected in biochemical systems, and illustrate the potential of this work for exploitation in the synthesis of amino acids and their derivatives.

ACKNOWLEDGMENTS

I gratefully acknowledge all the intellectual and practical contributions to our work in this area of Mr. T. Badran, Dr. V. Burgess, Mr. B. Clark, Dr. M. Hay, Mr. C. Hutton,

Mr. M. Kling, Mr. S. Love, Mr. S. Peters, Mr. M. Pitt, Mr. P. Roselt, Ms. G. Rositano, Ms. I. Scharfbillig, Dr. E. W. Tan and Ms. S. Watkins. I thank Dr. M. Crossley for disclosing results prior to publication and I am grateful to Ms. G. Rositano for assistance with the preparation of this chapter.

REFERENCES

1. Baldwin, J. E.; Abraham, E. P. *Nat. Prod. Rep.* **1988**, *5*, 129–145.
2. Baldwin, J. E.; Wan, T. S. *J. Chem. Soc., Chem. Commun.* **1979**, 249–250.
3. Baldwin, J. E.; Adlington, R. M.; Domayne-Hayman, B. P.; Knight, G.; Ting, H.-H. *J. Chem. Soc., Chem. Commun.* **1987**, 1661–1663.
4. Baldwin, J. E.; Adlington, R. M.; Derome, A. E.; Ting, H.-H.; Turner, N. J. *J. Chem. Soc., Chem. Commun.* **1984**, 1211–1214.
5. Baldwin, J. E.; Adlington, R. M.; Kang, T. W.; Lee, E.; Schofield, C. J. *J. Chem. Soc., Chem. Commun.* **1987**, 104–106.
6. Smith, K. C. *Photochem. Photobiol.* **1964**, *3*, 415–427.
7. Toth, B.; Dose, K. *Rad. Environ. Biophys.* **1976**, *13*, 105–113.
8. Dietz, T. M.; von Trebra, R. J.; Swanson, B. J.; Koch. T. D. *J. Am. Chem. Soc.* **1987**, *109*, 1793–1797.
9. Collins, M. A.; Grant, R. A. *Photochem. Photobiol.* **1969**, *9*, 369–375.
10. Easton, C. J.; Bowman, N. J. *J. Chem. Soc., Chem. Commun.* **1983**, 1193–1194.
11. Baldwin, J. E.; Adlington, R. M.; Bohlmann, R. *J. Chem. Soc., Chem. Commun.* **1985**, 357–359.
12. Barton, D. H. R.; Hervé, Y.; Potier, P.; Thierry, J. *Tetrahedron* **1987**, *43*, 4297–4308.
13. Barton, D. H. R.; Hervé, Y.; Potier, P.; Thierry, J. *Tetrahedron* **1988**, *44*, 5479–5486.
14. Barton, D. H. R.; Crich, D.; Hervé, Y.; Potier, P.; Thierry, J. *Tetrahedron* **1985**, *41*, 4347–4357.
15. Barton, D. H. R.; Bridon, D.; Hervé, Y.; Potier, P.; Thierry, J.; Zard, S. Z. *Tetrahedron* **1986**, *42*, 4983–4990.
16. Barton, D. H. R.; Guilhem, J.; Hervé, Y.; Potier, P.; Thierry, J. *Tetrahedron Lett.* **1987**, *28*, 1413–1416.
17. Baldwin, J. E.; Adlington, R. M.; Basak, A. *J. Chem. Soc., Chem. Commun.* **1984**, 1284–1285.
18. Baldwin, J. E.; Li, C.-S. *J. Chem. Soc., Chem. Commun.* **1987**, 166–168.
19. Easton, C. J.; Scharfbillig, I. M.; Tan, E. W. *Tetrahedron Lett.* **1988**, *29*, 1565–1568.
20. Baldwin, J. E.; Adlington, R. M.; Lowe, C.; O'Neil, I. A.; Sanders, G. L.; Schofield, C. J.; Sweeney, J. B. *J. Chem. Soc., Chem. Commun.* **1988**, 1030–1031.
21. Burgess, V. A.; Easton, C. J.; Hay, M. P.; Steel, P. J. *Aust. J. Chem.* **1988**, *41*, 701–710.
22. Burgess, V. A.; Easton, C. J. *Aust. J. Chem.* **1988**, *41*, 1063–1070.
23. Easton, C. J.; Tan, E. W.; Hay, M. P. *J. Chem. Soc., Chem. Commun.* **1989**, 385–386.
24. Easton, C. J.; Scharfbillig, I. M. *J. Org. Chem.* **1990**, *55*, 384–386.
25. Crich, D.; Davies, J. W.; Negrón, G.; Quintero, L.; *J. Chem. Res. (S)* **1988**, 140–141.
26. Orlinski, R.; Stankiewicz, T. *Tetrahedron Lett.* **1988**, *29*, 1601–1602.
27. Crich, D.; Davies, J. W. *Tetrahedron* **1989**, *45*, 5641–5654.
28. Wagner, I.; Musso, H. *Angew. Chem., Int. Ed. Eng.* **1983**, *22*, 816–828.
29. Schaich, K. M. *CRC Crit. Rev. Food Sci. Nutr.* **1980**, *13*, 131–159.
30. Henriksen, T.; Melø, T. B.; Saxebøl, G. In *Free Radicals in Biology*; Pryor, W. A., Ed.; Academic Press: New York, 1976; Vol. 2, pp 213–256.
31. Santus, R.; Grossweiner, L. I. *Photochem. Photobiol.* **1972**, *15*, 101–105.
32. Grossweiner, L. I.; Usui, Y. *Photochem. Photobiol.* **1970**, *11*, 53–56.

33. Androes, G. M.; Gloria, H. R.; Reinisch, R. F. *Photochem. Photobiol.* **1972**, *15*, 375–393.
34. Cowgill, R. W. *Biochim. Biophys. Acta* **1967**, *140*, 37–44.
35. Miyagawa, I.; Kurita, Y.; Gordy, W. *J. Chem. Phys.* **1960**, *33*, 1599–1603.
36. Katayama, M.; Gordy, W. *J. Chem. Phys.* **1961**, *35*, 117–122.
37. Rosenthal, I.; Poupko, R.; Elad, D. *J. Phys. Chem.* **1973**, *77*, 1944–1948.
38. Paul, H.; Fischer, H. *Ber. Bunsenges. Phys. Chem.* **1969**, *73*, 972–980.
39. Neta, P.; Fessenden, R. W. *J. Phys. Chem.* **1971**, *75*, 738–748.
40. Kirino, Y.; Taniguchi, H. *J. Am. Chem. Soc.* **1976**, *98*, 5089–5096.
41. Taniguchi, H.; Hatano, H.; Hasegawa, H.; Maruyama, T. *J. Phys. Chem.* **1970**, *74*, 3063–3065.
42. Livingston, R.; Doherty, D. G.; Zeldes, H. *J. Am. Chem. Soc.* **1975**, *97*, 3198–3204.
43. Viehe, H. G.; Merényi, R.; Stella, L.; Janousek, Z. *Angew. Chem., Int. Ed. Eng.* **1979**, *18*, 917–932.
44. Viehe, H. G.; Janousek, Z.; Merényi, R.; Stella, L. *Acc. Chem. Res.* **1985**, *18*, 148–154.
45. Dewar, M. J. S. *J. Am. Chem. Soc.* **1952**, *74*, 3353–3354.
46. Balaban, A. T. *Rev. Roum. Chim.* **1971**, *16*, 725–737.
47. Negoita, N.; Baican, R.; Balaban, A. T. *Tetrahedron Lett.* **1973**, 1877–1878.
48. Balaban, A. T.; Istratoiu, R. *Tetrahedron Lett.* **1973**, 1879–1880.
49. Baldock, R. W.; Hudson, P.; Katritzky, A. R.; Soti, F. *J. Chem. Soc., Perkin Trans. 1* **1974**, 1422–1427.
50. Katritzky, A. R.; Soti, F. *J. Chem. Soc., Perkin Trans. 1* **1974**, 1427–1432.
51. Katritzky, A. R.; Zerner, M. C.; Karelson, M. M. *J. Am. Chem. Soc.* **1986**, *108*, 7213–7214.
52. Pasto, D. J.; Krasnansky, R.; Zercher, C. *J. Org. Chem.* **1987**, *52*, 3062–3072.
53. Pasto, D. J. *J. Am. Chem. Soc.* **1988**, *110*, 8164–8175.
54. Korth, H.-G.; Lommes, P.; Sustmann, R. *J. Am. Chem. Soc.* **1984**, *106*, 663–668.
55. Korth, H.-G.; Lommes, P.; Sicking, W.; Sustmann, R. *Chem. Ber.* **1985**, *118*, 4627–4631.
56. MacInnes, I.; Walton, J. C.; Nonhebel, D. C. *J. Chem. Soc., Chem. Commun.* **1985**, 712–713.
57. Louw, R.; Bunk, J. *Recl.: J. R. Neth. Chem. Soc.* **1983**, *102*, 119–120.
58. Barbe, W.; Beckhaus, H.-D.; Rückhart, C. *Chem. Ber.* **1983**, *116*, 3216–3234.
59. Birkhofer, H.; Hádrich, J.; Beckhaus, H.-D.; Rückart, C. *Angew. Chem., Int. Ed. Eng.* **1987**, *26*, 573–575.
60. Burgess, V. A.; Easton, C. J. *Tetrahedron Lett.* **1987**, *28*, 2747–2750.
61. Taniguchi, H.; Kirino, Y. *J. Am. Chem. Soc.* **1977**, *99*, 3625–3631.
62. Moriya, F.; Makino, K.; Suzuki, N.; Rokushika, S.; Hatano, H. *J. Am. Chem. Soc.* **1982**, *104*, 830–836.
63. Poupko, R.; Silver, B. L.; Lowenstein, A. *J. Chem. Soc., Chem. Commun.* **1968**, 453.
64. Smith, P.; Fox, W. M.; McGinty, D. J.; Stevens, R. D. *Can. J. Chem.* **1970**, *48*, 480–491.
65. Elad, D.; Sperling, J. *J. Chem. Soc., Chem. Commun.* **1968**, 655–656.
66. Elad, D.; Sperling, J. *J. Chem. Soc., Chem. Commun.* **1969**, 234.
67. Elad, D.; Sperling, J. *J. Chem. Soc. C* **1969**, 1579–1585.
68. Elad, D.; Schwarzberg, M.; Sperling, J. *J. Chem. Soc., Chem. Commun.* **1970**, 617–618.
69. Sperling, J.; Elad, D. *J. Am. Chem. Soc.* **1971**, *93*, 967–971.
70. Sperling, J.; Elad, D. *J. Am. Chem. Soc.* **1971**, *93*, 3839–3840.
71. Schwarzberg, M.; Sperling, J.; Elad, D. *J. Am. Chem. Soc.* **1973**, *95*, 6418–6426.
72. Obata, N.; Niimura, K. *J. Chem. Soc., Chem. Commun.* **1977**, 238–239.
73. Lidert, Z.; Gronowitz, S. *Synthesis* **1980**, 322–324.
74. Kober, R.; Steglich, W. *Liebigs Ann. Chem.* **1983**, 599–609.
75. Kober, R.; Papadopoulos, K.; Miltz, W.; Enders, D.; Steglich, W.; Reuter, H.; Puff, H. *Tetrahedron* **1985**, *41*, 1693–1701.
76. Sinclair, P. J.; Zhai, D.; Reibenspeis, J.; Williams, R. M. *J. Am. Chem. Soc.* **1986**, *108*, 1103–1104.

77. Williams, R. M.; Zhai, D.; Sinclair, P. J.; *J. Org. Chem.* **1986**, *51*, 5021–5022.
78. Zimmermann, J.; Seebach, D. *Helv. Chim. Acta* **1987**, *70*, 1104–1114.
79. Williams, R. M.; Sinclair, P. J.; Zhai, D.; Chen, D. *J. Am. Chem. Soc.* **1988**, *110*, 1547–1557.
80. Ermert, P.; Meyer, J.; Stucki, C.; Schneebeli, J.; Obrecht, J.-P. *Tetrahedron Lett.* **1988**, *29*, 1265–1268.
81. Bowman, N. J.; Easton, C. J., University of Canterbury, New Zealand, unpublished observations.
82. Easton, C. J.; Peters, S. C., University of Adelaide, Australia, unpublished observations.
83. Wheelan, P.; Kirsch, W. M.; Koch, T. D. *J. Org. Chem.* **1989**, *54*, 4360–4364.
84. Easton, C. J.; Hay, M. P.; Love, S. G. *J. Chem. Soc., Perkin Trans. 2* **1988**, 265–268.
85. Easton, C. J.; Hay, M. P. *J. Chem. Soc., Chem. Commun.* **1986**, 55–57.
86. Burgess, V. A.; Easton, C. J.; Hay, M. P. *J. Am. Chem. Soc.* **1989**, *111*, 1047–1052.
87. Russell, G. A. In *Free Radicals*; Kochi, J. K., Ed.; Wiley: New York, 1973; Vol. 1, pp 275–331.
88. Shono, T.; Matsumura, Y.; Tsubata, K.; Sugihara, Y.; Yamane, S.; Kanazawa, T.; Aoki, T. *J. Am. Chem. Soc.* **1982**, *104*, 6697–6703.
89. Bradbury, A. F.; Finnie, M. D. A.; Smyth, D. G. *Nature* **1982**, *298*, 686–688.
90. Eipper, B. A.; Mains, R. E.; Glembotski, C. C. *Proc. Natl. Acad. Sci. U.S.A.* **1983**, *80*, 5144–5148.
91. Young, S. D.; Tamburini, P. P. *J. Am. Chem. Soc.* **1989**, *111*, 1933–1934.
92. Davies, J. S.; Thomas, R. J.; Williams, M. K. *J. Chem. Soc., Chem. Commun.* **1975**, 76–77.
93. Davies, J. S.; Thomas, R. J. *J. Chem. Soc., Perkin Trans. 1* **1981**, 1639–1646.
94. Bowman, N. J.; Hay, M. P.; Love, S. G.; Easton, C. J. *J. Chem. Soc., Perkin Trans. 1* **1988**, 259–264.
95. Burgess, V. A., Ph.D. Dissertation, University of Adelaide, Adelaide, Australia, 1989.
96. Liesch, J. M.; Millington, D. S.; Pandey, R. C.; Rinehart, K. L. *J. Am. Chem. Soc.* **1976**, *98*, 8237–8249.
97. Ito, Y.; Ohashi, Y.; Kawabe, S.; Abe, H.; Okuda, T. *J. Antibiot.* **1972**, *25*, 360–361.
98. Konishi, M.; Ohkuma, H.; Sakai, F.; Tsuno, T.; Koshiyama, H.; Naito, T.; Kawaguchi, H. *J. Am. Chem. Soc.* **1981**, *103*, 1241–1243.
99. Arnold, E.; Clardy, J. *J. Am. Chem. Soc.* **1981**, *103*, 1243–1244.
100. Clark, B. M., M.Sc. Dissertation, University of Canterbury, Christchurch, New Zealand, 1987.
101. Johnson, R. A.; Greene, F. D. *J. Org. Chem.* **1975**, *40*, 2186–2192.
102. Barton, D. H. R.; Beckwith, A. L. J.; Goosen, A. *J. Chem. Soc.* **1965**, 181–190.
103. Beckwith, A. L. J.; Goodrich, J. E. *Aust. J. Chem.* **1965**, *18*, 747–757.
104. Neale, R. S.; Marcus, N. L.; Schepers, R. G. *J. Am. Chem. Soc.* **1966**, *88*, 3051–3058.
105. Chow, Y. L.; Joseph, T. C. *J. Chem. Soc., Chem. Commun.* **1969**, 490–491.
106. Clark, B. M.; Easton, C. J., unpublished observations.
107. Sperling, J.; Elad, D. *Isr. J. Chem.* **1967**, *5*, 40p.
108. Hausler, J.; Jahn, R.; Schmidt, U. *Chem. Ber.* **1978**, *111*, 361–368.
109. Coyle, J. D.; Hill, R. R.; Randall, D.; *Photochem. Photobiol.* **1984**, *40*, 153–159.
110. Trown, P. W. *Biochem. Biophys. Res. Commun.* **1968**, *33*, 402–407.
111. Yoshimura, J.; Nakamura, H.; Matsunari, K. *Bull. Chem. Soc. Jpn.* **1975**, *48*, 605–609.
112. Nakatsuka, S.; Yamada, K.; Yoshida, K.; Asano, O.; Murakami, Y.; Goto, T. *Tetrahedron Lett.* **1983**, *24*, 5627–5630.
113. Shimazaki, N.; Shima, I.; Hemmi, K.; Tsurumi, Y.; Hashimoto, M. *Chem. Pharm. Bull.* **1987**, *35*, 3527–3530.
114. Williams, R. M.; Kwast, A. *J. Org. Chem.* **1988**, *53*, 5785–5787.
115. Badran, T. W.; Easton, C. J. *Aust. J. Chem.* **1990**, *43*, 1455–1459.
116. Badran, T. W.; Easton, C. J., University of Adelaide, Australia, unpublished observations.
117. Easton, C. J.; Scharfbillig, I. M., University of Adelaide, Australia, unpublished observations.

118. Easton, C. J.; Hay, M. P. *J. Chem. Soc., Chem. Commun.* **1985**, 425–427.
119. Beckwith, A. L. J.; Ingold, K. U. In *Rearrangements in Ground and Excited States*; de Mayo, P., Ed.; Academic Press: New York, **1980**; Vol. 1, pp 161–310.
120. Kuivila, H. G. *Acc. Chem. Res.* **1968**, *1*, 299–305.
121. Easton, C. J.; Findlay, N. G. *J. Lab. Comp. Radiopharm.* **1984**, *22*, 667–672.
122. Keck, G. E.; Yates, J. B. *J. Organomet. Chem.* **1983**, *248*, C21–C25.
123. Keck, G. E.; Enholm, E. J.; Yates, J. B.; Wiley, M. R. *Tetrahedron* **1985**, *41*, 4079–4094.
124. Grignon, J.; Pereyre, M. *J. Organomet. Chem.* **1973**, *61*, C33–C35.
125. Servens, C.; Pereyre, M. *J. Organomet. Chem.* **1971**, *26*, C4–C6.
126. Khoo, L. E.; Lee, H. H. *Tetrahedron Lett.* **1968**, 4351–4354.
127. Giese, B. In *Radicals in Organic Synthesis: Formation of Carbon–Carbon Bonds*; Pergamon Press: New York, **1986**.
128. Crossley, M. J.; Reid, R. C., University of Sydney, Australia, personal communication.
129. Dowd, P.; Choi, S.-C.; Duah, F.; Kaufman, C. *Tetrahedron* **1988**, *44*, 2137–2148.
130. Choi, S.-C.; Dowd, P. *J. Am. Chem. Soc.* **1989**, *111*, 2313–2314.
131. Barker, H. A.; Weissbach, H.; Smyth, R. D. *Proc. Natl. Acad. Sci. U.S.A.* **1958**, *44*, 1093–1097.
132. Seebach, D.; Colvin, E. W.; Lehr, F.; Weller, T. *Chimia* **1979**, *33*, 1–18.
133. Seebach, D.; Lehr, F. *Angew. Chem., Int. Ed. Eng.* **1976**, *15*, 505–506.
134. Seebach, D.; Henning, R.; Lehr, F.; Gonnermann, J. *Tetrahedron Lett.* **1977**, *18*, 1161–1164.
135. Kerber, R. C.; Urry, G. W.; Kornblum, N. *J. Am. Chem. Soc.* **1964**, *86*, 3904–3905.
136. Kornblum, N.; Boyd, S. D.; Ono, N. *J. Am. Chem. Soc.* **1974**, *96*, 2580–2587.
137. Castelhano, A. L.; Horne, S.; Billedeau, R.; Krantz, A. *Tetrahedron Lett.* **1986**, *27*, 2435–2438.
138. Munster, P.; Steglich, W. *Synthesis* **1987**, 223–225.
139. Koch, T. H.; Olesen, J. A.; DeNiro, J. *J. Am. Chem. Soc.* **1975**, *97*, 7285–7288.
140. Koch, T. H.; Olesen, J. A.; DeNiro, J. *J. Org. Chem.* **1975**, *40*, 14–19.
141. Kleyer, D. L.; Haltiwanger, R. C.; Koch, T. H. *J. Org. Chem.* **1983**, *48*, 147–152.
142. Olson, J. B.; Koch, T. H. *J. Am. Chem. Soc.* **1986**, *108*, 756–761.
143. Bennett, R. W.; Wharry, D. L.; Koch, T. H. *J. Am. Chem. Soc.* **1980**, *102*, 2345–2349.
144. Burns, J. M.; Wharry, D. L.; Koch, T. H. *J. Am. Chem. Soc.* **1981**, *103*, 849–856.
145. Kleyer, D. L.; Koch, T. H. *J. Am. Chem. Soc.* **1983**, *105*, 5154–5155.
146. Kleyer, D. L.; Koch, T. H. *J. Am. Chem. Soc.* **1983**, *105*, 5911–5912.
147. Kleyer, D. L.; Gaudiano, G.; Koch, T. H. *J. Am. Chem. Soc.* **1984**, *106*, 1105–1109.
148. Kleyer, D. L.; Koch, T. H. *J. Am. Chem. Soc.* **1984**, *106*, 2380–2387.
149. Bird, D. M.; Boldt, M.; Koch, T. H. *J. Am. Chem. Soc.* **1987**, *109*, 4046–4053.

CYCLOADDITIONS OF ALLENES

REACTIONS OF UNUSUAL

MECHANISTIC PERSPICUITY

William R. Dolbier, Jr.

Advances in Detailed Reaction Mechanisms
Volume 1, pages 127–179
Copyright © 1991 JAI Press Inc.
All rights of reproduction in any form reserved.
ISBN: 1-55938-164-7

1. INTRODUCTION

Cycloadditions are a class of reactions that only thirty years ago were for the most part considered mechanistically obscure and primarily of interest to physical organic chemists. The physical organic chemists themselves found these so-called "no-mechanism" reactions intellectually fascinating, but usually experimentally inscrutable. In contrast, today the high level of understanding of all aspects of cycloaddition reactions, and their acknowledged unique regiochemical and stereochemical characteristics, make them among the most highly used and indispensable tools of the practicing chemist. Nevertheless, challenges remain in our quest to elucidate the substantial remaining mechanistic mysteries and to understand more fully the variable structure/reactivity relationships that underlie the various types of cycloaddition reactions.

Cycloadditions are processes in which two or more reactants combine to form a stable cyclic molecule; during the process σ bonds are formed at the expense of π bonds and no small fragments are eliminated. Such reactions may be pericyclic in nature (that is, they may be "reactions in which all first order changes in bonding relationships take place in concert on a closed curve")[1,2] or they may be reactions in which the two σ bonds are formed stepwise, that is, via a first σ-bond-forming step to produce a transient intermediate, usually a diradical, that cyclizes in a second step to form the product.

Indeed, that is the central question that physical organic chemists have from the beginning tried to answer experimentally: whether cycloaddition reactions are concerted in nature or whether they proceed via discrete intermediates. Application of the usual kinetic and nonkinetic probes of mechanism does not allow one to prove conclusively that a reaction is concerted; one can merely eliminate reasonable alternatives until he or she can "zero in" on the correct mechanism.

This raises the basic question of how one *does* in general determine the details of a mechanism; more specifically, how does one demonstrate the existence of intermediates in a reaction? First of all one needs to identify the products, including minor side products, if present. (These often provide vital clues to the existence and the nature of intermediates.) Then one looks for direct evidence of an intermediate: trapping it or, more commonly, simply detecting it indirectly, such as by demonstrating kinetically that the rate-determining and product-determining steps are not the same step. Other experimental techniques that may provide insight into the intervention of an intermediate include the observation of scrambling of an isotopic label; studies of the regiochemistry and the stereochemistry of the reaction; rate studies; studies of solvent effects; isotope effects; and effects of substituents on reaction rates. By their nature, these techniques are excellent for proving that there *is* an intermediate, but not so good for proving that there is *not* an intermediate. A good example of that is

$$CF_2=CCl_2 \quad + \qquad \xrightarrow{\Delta,\ 80°}$$

Bartlett's classic stereochemical study of [2 + 2] reactions,[3,4] which resulted in virtually every organic chemist being convinced by 1965 that such reactions are two-step, and that they involve diradical intermediates.

In contrast, with regard to the mechanisms of Diels–Alder and 1,3-dipolar cycloadditions, there was still considerable controversy in 1965 as to whether these reactions were indeed concerted, although the preponderance of opinion then would have been that they were. There just was not much hard evidence for these convictions. Enter Woodward and Hoffmann with their principles of orbital symmetry, which indicated that both the Diels–Alder reaction and 1,3-dipolar cycloadditions were *allowed* to take place in a concerted manner, while [2 + 2] cycloadditions were *not* allowed to take place via a concerted mechanism.[1,2] Woodward and Hoffmann, of course, addressed many other potentially pericyclic reactions in their classic series of papers, and over the last twenty-five years the efficacy of these predictive rules has been borne out virtually without exception through experiment. It must be said, however, that a mechanism has never been *proven* by the invocation of a rule. Although Woodward and Hoffmann's intellectually satisfying rules certainly gave one greater confidence in the assumption that Diels–Alder and 1,3-dipolar cycloadditions were, as classes of reactions, concerted, there was still the need to develop new and more definitive mechanistic probes that could continue to build the experimental basis for this conclusion.

At about the same time, a not-unrelated controversy was brewing regarding the mechanism of allene [2 + 2] cycloadditions. A number of kinetic[5] and stereochemical[6–12] studies had been carried out, the results of which suggested that allene [2 + 2] cycloadditions might well be taking place via a multicenter, concerted mechanism, with allene acting in an antarafacial manner similar to the way that ketene would seem to be behaving.[13–17] Especially interesting was the work by Moore,[7] wherein the results from dimerization of both racemic and optically active 1,2-cyclononadiene are consistent with the process proceeding

a single enantiomer   meso only

apparently via a
suprafacial-antarafacial
[2 + 2] cycloaddition

via a suprafacial–antarafacial [2 + 2] cycloaddition, in agreement with Woodward–Hoffmann predictions for a concerted cycloaddition.[2] Nevertheless, our feeling at that time was that—while allene was indeed a unique [2 + 2] addend the behavior of which could be used to considerable advantage in probing mechanism (as we will see)—it was unlikely that allene would participate in any such strange mechanism. Moreover, the unusual stereochemical observations could be rationalized just as reasonably by the invocation of diradical intermediates, as indeed Moore realized.

1.1. Allenes as Addends in Cycloaddition Reactions

Allenes are cumulated dienes that thus have two contiguous, orthogonal π bonds. A significant "strain" is found to be associated with allenes (about 10–11 kcal/mol, as determined from relative heats of hydrogenation)[18]; thus some enhancement of reactivity is expected owing to this strain. However, unactivated allenes (that is, parent and alkyl-substituted allenes) are found to be relatively unreactive in normal Diels–Alder and 1,3-dipolar cycloadditions. In fact allene behaves like a relatively electron-rich alkene, reacting much more readily with hexachlorocyclopentadiene than with cyclopentadiene.[19,20] This finding should not have been surprising since, based upon adiabatic ionization potentials, allene (9.69 eV) is about as electron-rich as propene (9.75 eV).[21]

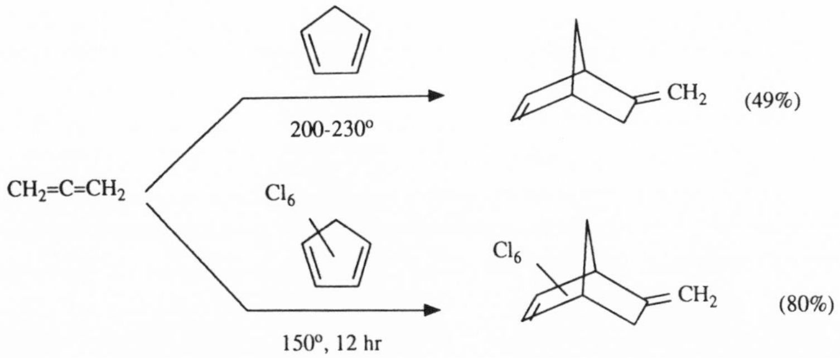

Allene is, however, quite reactive in undergoing [2 + 2] cycloadditions.[22] While the strain factor certainly may contribute to such reactivity, much more important is the fact that initial bond formation in such reactions occurs at C_2 of the allene, and concomitant with this bond formation occurs a rotation of the

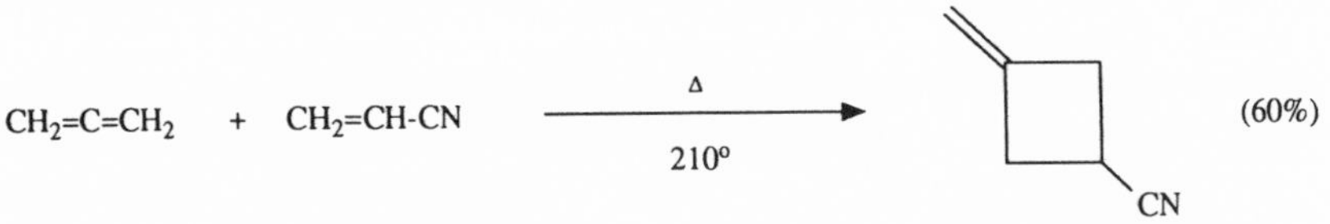

terminus of the reactive bond 90° to become coplanar with the orthogonal double bond, thus forming a stabilized allyl radical. With the transition state of this step expected to be very product-like, it should thus be very allyl-radical-like and therefore quite stabilized.

The fact is, however, that allene *is* sufficiently reactive to act as an addend in undergoing cycloadditions of all of the classic types: Diels–Alder, 1,3-dipolar, and [2 + 2] cycloaddition. Recognizing the unique symmetry characteristics of allene, i.e., its having two identical, noninteracting double bonds, in 1968 we identified allene cycloadditions as an ideal system within which to probe mechanism.

1.2. Possible Mechanisms

The Concerted Mechanism

In a concerted process, the other addend, such as a diene or a 1,3-dipole, would have to choose, in the one and only, rate-determining and product-determining step, with which of allene's two bonds to react. In allene itself, of course, the two π bonds would be identical, but in any unsymmetrically substituted allene the reactant would have to make a choice, and the factors that would be involved in making this choice would be very mechanism-dependent.

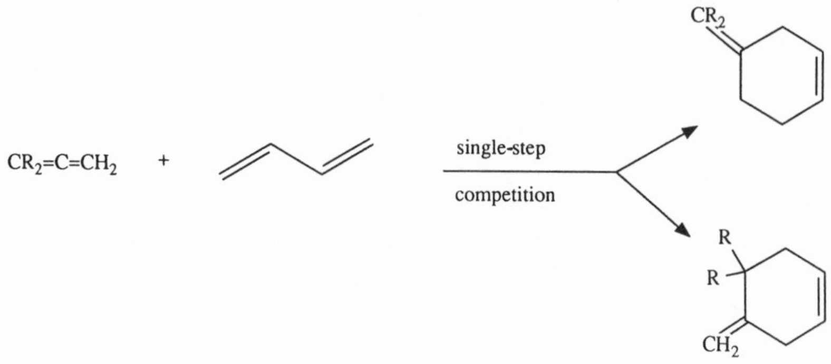

Perturbational molecular orbital theory predicts that the rates of concerted cycloadditions, such as Diels–Alder and 1,3-dipolar cycloadditions of allenes, should be largely dependent upon the relative energies of the frontier molecular orbitals of the dienes or dipoles and those of the dienophilic or dipolarophilic allene.[23,24] In the normal case, this means that the rates of such allene cycloadditions should be almost wholly dependent upon the energy of the LUMOs of the allene addends—the *lower* the energy of the LUMO, the more reactive the allene. With allene's symmetry, its HOMO and LUMO molecular orbitals are both degenerate, and its two double bonds should be equally reactive. Unsymmetrical substitution of allene would, of course, destroy the degeneracies of the π and π^* molecular orbitals.

The Two-Step Mechanism

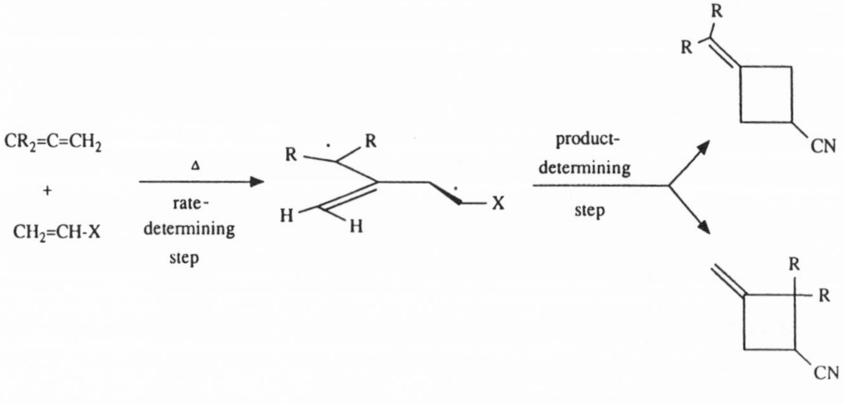

In a two-step reaction, the nonallene addend would react with one or the other allene double bond, always at C_2, to form an intermediate allyl-stabilized diradical, which would then cyclize to one or the other end of the allyl system. Thus the product-determining step in the nonconcerted process would involve a presumably low-activation-energy rotation of one end of the allyl system out of conjugation to form the second σ bond, a regiochemistry-determining process that is certainly very different from that in the concerted reaction.

In the concerted reaction, the product-determining decision is in choosing *which* π bond to react with. In the stepwise process, this choice is immaterial since the same diradical intermediate will be obtained from reaction with either of the π bonds. The product-determining step in this case is the second step, in which a choice must be made as to which end of the allyl radical will undergo cyclization. For each mechanism the rate-determining step would be the same, but the two processes would have different product-determining steps. The net result is that the choice of which double bond ultimately has undergone reaction is made very differently for the concerted and the nonconcerted processes.

The potential was thus there to use *regioselectivity* of cycloadditions of unsymmetrically substituted allenes to probe the simultaneity of bond formation in cycloaddition reactions. One had merely to design the appropriately substituted allene systems to do the job.

2. CYCLOADDITIONS OF DEUTERATED ALLENES: A SECONDARY DEUTERIUM ISOTOPE EFFECT STUDY

It was decided initially to utilize deuterium substitution as a very subtle perturbation of the symmetry of the allene system in order to probe the diversity of allene cycloaddition mechanisms, and 1,1-dideuterioallene (DDA) was our

substrate of choice for the study. In addition to determining the relative reactivity of the CH_2 versus the CD_2 ends of the unsymmetrical 1,1-dideuterioallene in [2 + 2] and [2 + 4] cycloadditions, we also proposed to compare the reactivity of allene and tetradeuterioallene in such reactions.

2.1. Precedent and Predictions

Perturbation of the symmetry of allene by such deuterium substitution should, of course, not significantly modify allene's orbital degeneracy, and, as such, it was believed that isotope effect studies using 1,1-dideuterioallene would provide considerable insight into the mechanisms of allene [2 + 4] and [2 + 2] cycloadditions. Much was already known at that time about the kinetic effect of deuterium substitution on reactions in which changes in carbon hybridization occurred.[25-29]

Understanding such precedent, it was our hypothesis that under the conditions of a concerted cycloaddition the change in hybridization of a terminal CH_2 versus a terminal CD_2 would be the primary factor that would determine these relative rates, and precedent indicated that an *inverse* isotope effect ($k_H/k_D < 1$) should always be observed for a process in which carbon hybridization is modified from sp^2 to sp^3.[25-29]

At that time we were not too sure what kind of isotope effect to expect for a nonconcerted mechanism. Evidence on radical additions to isotopically labeled alkenes indicated that there should be little, if any, kinetic isotope effect for the addition step,[30] and there were virtually no available data on isotope effects for the second, fast cyclization step. (Our work with DDA would in the end provide the first such data, primarily because we would be looking at *intramolecular*

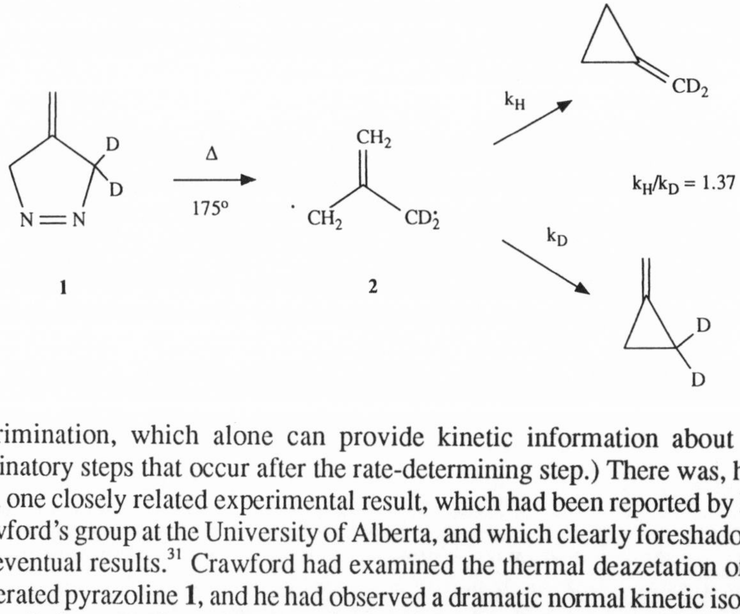

discrimination, which alone can provide kinetic information about discriminatory steps that occur after the rate-determining step.) There was, however, one closely related experimental result, which had been reported by R. J. Crawford's group at the University of Alberta, and which clearly foreshadowed our eventual results.[31] Crawford had examined the thermal deazetation of the deuterated pyrazoline 1, and he had observed a dramatic normal kinetic isotope effect ($k_H/k_D > 1$) for the cyclization of the expected trimethylenemethane intermediate 2. One could expect that the transition state for the rapid cyclization step in our hypothetical two-step, diradical mechanism for allene cycloaddition might resemble very closely that for cyclization of Crawford's trimethylenemethane intermediate.

Therefore, the predictions were as follows: In a concerted cycloaddition allene-d_4 should undergo cycloaddition faster than allene, and 1,1-dideuteroallene should undergo cycloaddition at a similarly greater rate with its deuterated

Predictions for Intramolecular Isotope Effect Study:

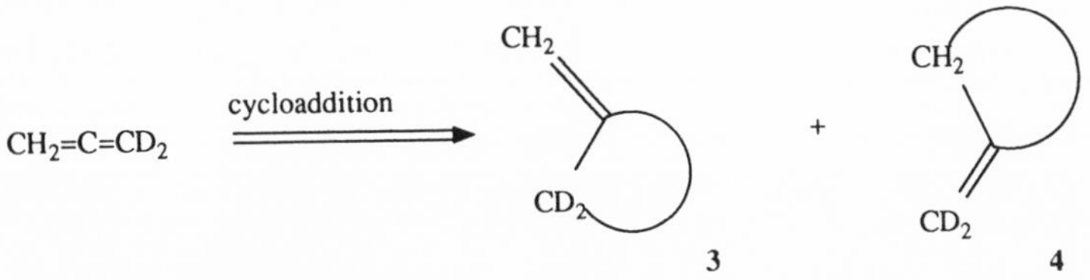

for concerted reaction: **3 > 4**

for nonconcerted reaction: **4 > 3**

double bond than with its nondeuterated double bond (i.e., more product **3** than product **4**). In a nonconcerted cycloaddition, all of the allenes will react at about the same rate, but the 1,1-dideuterioallene will in the end have undergone greater cyclization to its nondeuterated double bond (i.e., forming more product **4** than product **3**).

2.2. Cycloaddition Results

This was one of those rare projects in which the results turned out to be just as good as the predictions. A total of two Diels–Alder reactions, one 1,3-dipolar cycloaddition, and three [2 + 2] cycloadditions were studied.[32–35] The Diels–Alder reaction between allene and hexachlorocyclopentadiene (**5**) and the [2 + 2] cycloaddition reaction between allene and acrylonitrile (**6**) have been used to exemplify the type of intramolecular and intermolecular competition experiments that were carried out in order to determine the kinetic secondary deuterium isotope effects. The intramolecular isotope effects were determined

Intramolecular Isotope Effect Studies:

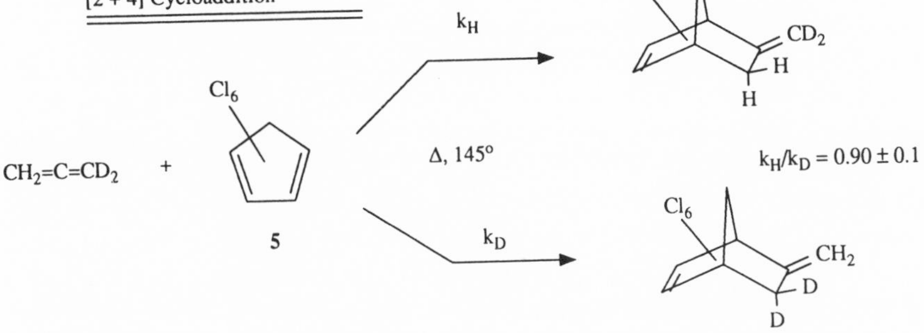

by NMR integration of vinylic versus allylic protons, while the intermolecular isotope effects were obtained by low-voltage mass spectrometric determination of the d_0 versus d_4 adducts.

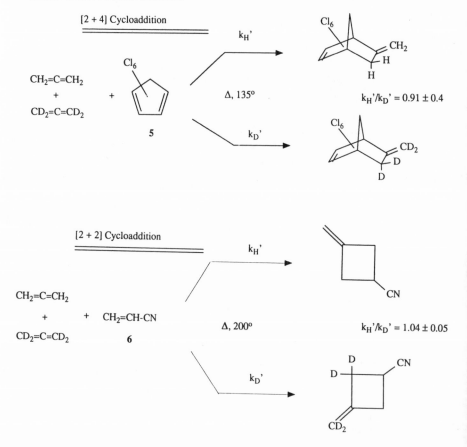

Intermolecular Isotope Effect Studies:

Diels–Alder Reaction: Identity of Rate-Determining and Product-Determining Steps

There can be little doubt regarding the basic conclusions to be drawn from the results given above. In the Diels–Alder reaction between allene and hexachlorocyclopentadiene virtually identical inverse intramolecular and intermolecular kinetic secondary deuterium isotope effects were obtained. This means that, for this reaction, the isotope effect for the *rate-determining* step is the same as the isotope effect for the *product-determining* step. If the reaction is concerted, then the rate-determining and product-determining steps would by necessity be *the same step*, and thus they would have identical kinetic isotope effects. Therefore the results are consistent with a concerted mechanism being operative in this [2 + 4] cycloaddition.

[2 + 2] Cycloaddition: Nonidentity of Rate-Determining and Product-Determining Steps

In contrast, by the same token, the results from the [2 + 2] cycloaddition of allene with acrylonitrile (6) effectively *preclude* a concerted reaction. That is, distinctly different intra- and intermolecular kinetic isotope effects were observed, with the intramolecular competition giving rise to a large normal isotope effect ($k_H/k_D = 1.21$), while only an experimentally negligible isotope effect ($k_H'/k_D' = 1.03 \pm 0.04$) is observed in the case of intermolecular competition. Note that the observed large normal *kinetic* isotope effect is in distinct contrast to the significant inverse *thermodynamic* isotope effect that can be observed by thermal equilibration of the kinetically formed adducts:

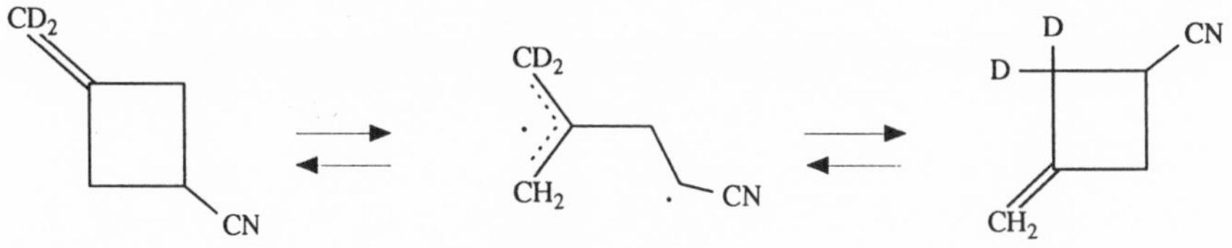

Δ, 280° for 15 hours

$k_H/k_D(\text{equil}) = 0.93 \pm 0.01$

Other Examples

Intramolecular isotope effects for the 1,3-dipolar cycloaddition between DDA and the carbonyl ylide (8) derived from tetracyanoethylene oxide (7) are consistent with this reaction also being a concerted one ($k_H/k_D = 0.93$), while another [2 + 2] reaction, that of DDA with 1,1-dichloro-2,2-

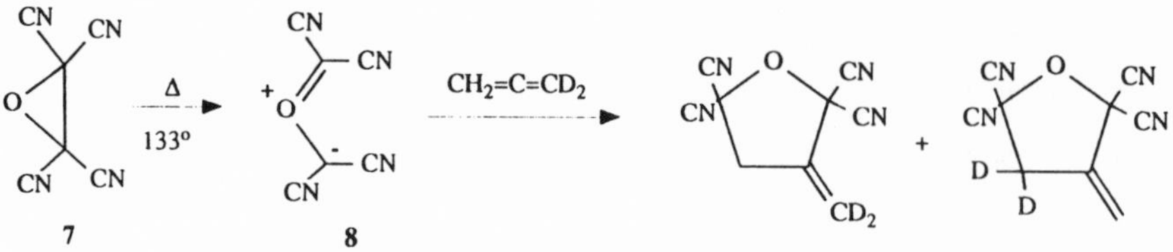

$k_H/k_D = 0.93 \pm 0.01$

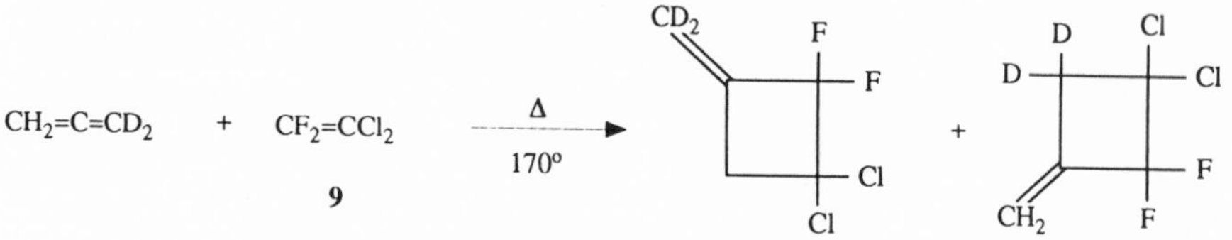

$k_H/k_D = 1.15 \pm 0.04$

difluoroethylene (9), clearly falls into the nonconcerted category, with k_H/k_D = 1.15.

Allene Dimerization:

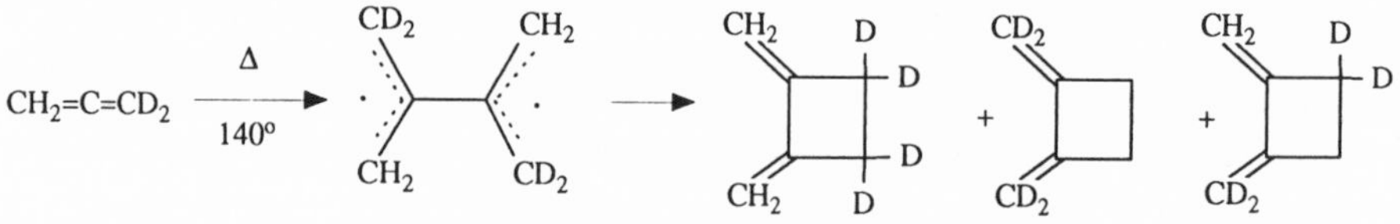

intramolecular $k_H/k_D > 1.14 \pm 0.01$

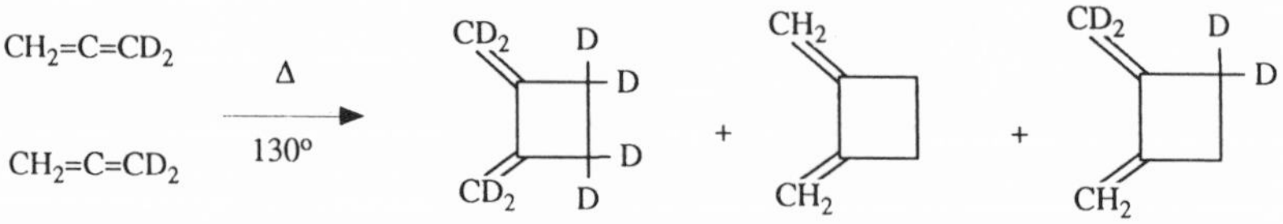

intermolecular $k_H'/k_D' = 1.02 \pm 0.03$

An important final study dealt with the especially significant case of the dimerization of allene. After all, it was in the case of an allene dimerization that the strongest case had been made for a possible concerted mechanism.[6]

However, as had been the case for the other [2 + 2] cycloadditions, a comparison of the observed intramolecular isotope effect with the observed intermolecular isotope effect effectively precluded the possibility of a concerted mechanism being involved. Indeed the results were totally consistent with the dimerization being just another [2 + 2] allene cycloaddition.

A Summing Up

The overall conclusion that must be reached from this study is that the mechanisms of allene Diels–Alder and 1,3-dipolar reactions are, of necessity, *different* from the mechanism of its [2 + 2] cycloadditions. More specifically, the [2 + 2] mechanism must be multistep, since the rate-determining and product-determining steps are constrained to be different steps. Although the ancillary conclusion that the Diels–Alder and 1,3-dipolar cycloadditions are thereby concerted is not absolutely required by the data, it is the most reasonable conclusion based upon the identity of the inter- and intramolecular isotope effects.

This study gave a clear indication that allene, in its cycloadditions, behaves simply as a reasonably normal electron-rich alkene, undergoing synchronous Diels–Alder and 1,3-dipolar cycloadditions, and multistep [2 + 2] cycloadditions. Moreover, this study clearly demonstrated the efficacy of secondary deuterium kinetic isotope effects as a sensitive and definitive probe into the mechanisms of allene cycloaddition reactions, as well as the particularly advantageous use of umsymmetrically substituted allenes in mechanistic studies of cycloadditions.

Parenthetically, it should be mentioned that others have also recognized the great potential of allene cycloadditions to provide mechanistic insight. In particular, the group led by Dan Pasto at Notre Dame has made many important contributions to this field that complement our work nicely, some of which will be discussed in this chapter.

Pasto's contributions in this area date from theoretical work that he published in 1979,[36] wherein he proposed a scenario through which our isotope effect results could be made consistent with an extraordinary concerted mechanism. As indicated earlier, just as mechanisms can not be proven by the invocation of a rule, so they cannot be disproven by a calculation. Indeed, subsequent work by Pasto's group,[37–39] including secondary deuterium isotope effect studies on allene [2 + 2] cycloadditions, led to their conclusion that allene [2 + 2] cycloadditions are nonconcerted and proceed via diradical intermediates.

3. CYCLOADDITIONS OF FLUORINATED ALLENES

3.1. Rationale and the Birth of an Idea

Although we believed that our isotope effect studies using deuterated allenes allowed virtually definitive mechanistic distinction between concerted and nonconcerted pathways, organic chemists as a group have always been reluctant to ascribe great and/or definitive mechanistic significance to such small effects. Thus we decided to move away from the use of isotope effects as our primary mechanistic probe.

Isotope effects of course derive from differences in transition state energy of but hundreds of *calories*. It is well known—and was demonstrated again in the equilibrium isotope effect study discussed earlier—that the inverse isotope effects observed in concerted cycloadditions derive from the fact that in a conversion of an sp^2 to an sp^3 hybridized carbon, there is a small thermodynamic advantage for a CD_2 group versus a CH_2 group. In considering how we might come up with a more dramatic mechanistic probe that might successfully command the attention of the physical organic community, it occurred to us that a similar conversion of a CF_2 group from sp^2 to sp^3 hybridization had been projected to carry with it a thermodynamic benefit of about 10 *kilo*calories.[40,41] It seemed, moreover, that no one had yet determined whether this dramatic thermodynamic factor had any kinetic repercusions.

We concluded that a study of the cycloadditions of 1,1-difluoroallene (DFA) might well provide significant mechanistic insight, for much the same reasons that the isotope effect study using 1,1-dideuterioallene had proved so effica-

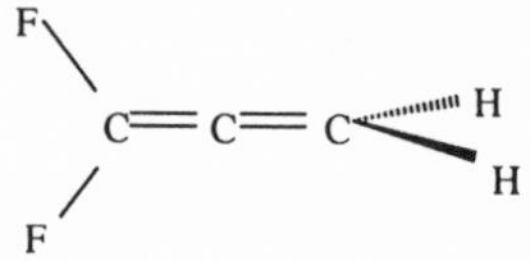

cious. It seemed reasonable that, with the involvement of such a strong thermodynamic factor in this system, concerted cycloadditions should favor addition to the fluorinated double bond, while a nonconcerted process should lead to ambiguous but certainly less regioselective results.

Imagine our surprise then when we found an early *Journal of the American Chemical Society* communication by Knoth and Coffman wherein it was claimed that cyclopentadiene underwent regiospecific Diels–Alder cycload-

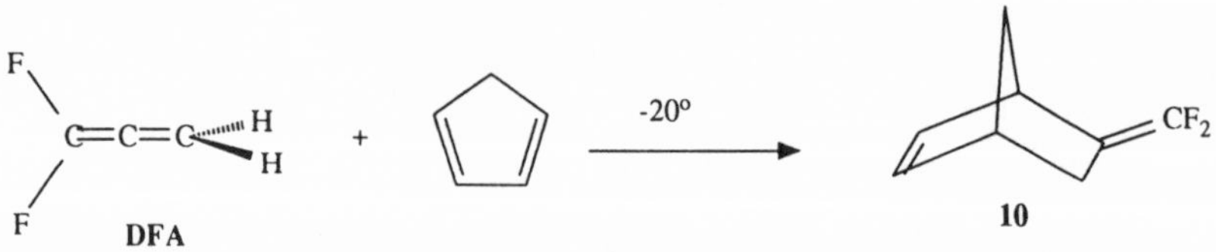

dition to the *non*fluorinated double bond of DFA to produce adduct **10**.[42] Since the report contained few data, we were somewhat skeptical and rushed to repeat the reaction, only to find that indeed the reaction proceeded as reported. (This was but the first of many erroneous predictions made by a hydrocarbon chemist in predicting the outcome of reactions of fluorine-containing hydrocarbons. Indeed, it was just this element of predictive challenge that made our excursion into the world of fluorine chemistry all the more attractive.)

Before proceeding further with our discussion of the cycloadditions of fluoroallenes, so that one might better understand the behavior of fluorine as a substituent in such reactions, it might be wise to discuss briefly the nature of fluorine as a substituent.

3.2. The Nature of Fluorine as a Substituent

Fluorine has many characteristics that serve to make it unique as a substituent. These characteristics are discussed in Smart's excellent reviews.[43,44] Fluorine is both the smallest and the most electronegative atomic substituent, and it is the best π donor of the halogen substituents. The latter characteristic is due to the fact that fluorine is a second-period element, and thus its atomic orbitals are of the ideal size for effective overlap with carbon orbitals both in forming σ bonds and in π-conjugative interaction with contiguous carbon π systems. Its high electronegativity and effective orbital overlap combine to result in a C–F σ bond that is both strong and short (1.385 Å for the CH_3–F bond versus 1.782 Å for the CH_3–Cl bond).[45] In fact, the fluorine substituent is the smallest of all nonhydrogen substituents, having an *A* value of 0.11 kcal as compared, for example, to *A* values of 1.8 for a methyl substituent and 0.7 for a chlorine substituent.[46] Examples of single fluorine substituents exerting a steric influence on the outcome of a reaction are almost nonexistent.

Fluorine's potentially strong electronic influences combined with its negligible size, plus the fact that it is an NMR-active nucleus with a very broad range of chemical shifts, make it unique as a substituent and particularly effective for use in probing mechanism.

Another factor, thermodynamic in nature, also plays a significant role in determining the reactivity-enhancing effect of fluorine on an olefin undergoing a cycloaddition reaction. It appears that the strength of C–F bonds is strongly dependent upon carbon's hybridization as well as upon the degree of accumulation of fluorine on a carbon atom. This effect is exemplified by the two experimental equilibria and two isodesmic equations represented at the top of the next page.[47]

As is the case for essentially all substituents, it can be seen that a *single* fluorine substituent apparently stabilizes a double bond [equation (1)], while surprisingly there is an observed, not inconsiderable thermodynamic advantage for geminal fluorine substituents (CF_2) to be located on an sp^3-hybridized carbon as opposed to an sp^2-hybridized carbon [equation (2)]. This likely derives from the large incremental geminal stabilization (IGSTAB) [equation (eq. 1) $FCH_2\text{-}CH=CH_2$ ⇌ $CHF=CH\text{-}CH_3$ (cis) ⇌ $CHF=CH\text{-}CH_3$ (trans)

$$\Delta H = -3.3 \text{ kcal/mol} \qquad\qquad \Delta H = +0.64 \text{ kcal/mol}$$

(eq. 2) $CH_2=CH\text{-}CHF_2$ $\underset{}{\overset{\Delta,\ I_2}{\rightleftarrows}}$ $CH_3\text{-}CH=CF_2$

$$\Delta H = +2.5 \text{ kcal/mol}$$

(eq. 3) $2\ CH_3CH_2F$ $\xrightarrow[\text{IGSTAB} = -6.6]{\Delta H = -13.1}$ CH_3CHF_2 + CH_3CH_3

$2(-62.9)^{11}$ $\qquad\qquad\qquad\qquad\qquad\qquad$ $(-118.8)^{12}$ $\qquad$ $(-20.1)^{12}$

(eq. 4) $2\ CH_2=CHF$ $\xrightarrow[\text{IGSTAB} = -2.5]{\Delta H = -5.0}$ $CH_2=CF_2$ + $CH_2=CH_2$

$2(-32.1)^{13}$ $\qquad\qquad\qquad\qquad\qquad\qquad$ $(-81.7)^{13}$ $\qquad$ $(+12.5)^{12}$

(3)] that is observed for saturated CF_2 but not for unsaturated CF_2 [equation (4)]. IGSTAB is the increase in thermodynamic stabilization of a geminal-substituted system, per substituent, relative to the respective monosubstituted system. Fluorine exhibits the largest IGSTAB on a saturated carbon of any substituent.[47]

3.3. Fluorine-Substituted Allenes

If one had but considered the nature of the Diels–Alder mechanism, as well as the expected effect of fluorine substituents on a π system such as allene, one would have been able to predict correctly the outcome of Knoth's cyclopentadiene reaction, even without much knowledge of fluorine chemistry. While most reactions, including cycloadditions, when given the choice do tend to lead to the more stable products, it was well known, as discussed earlier, that in Diels–Alder reactions the rate and the outcome of the reaction are controlled by the relative energies of the frontier molecular orbitals, that is, the HOMO of the diene and the LUMO of the dienophile.[48,49] Thus the better dienophile will be the one that is more electron deficient, that is, the one that has the lowest-energy LUMO π orbital.

Ab initio calculations have been carried out on both fluoroallene (MFA) and DFA, and these calculations indicate clearly that their LUMOs are their $C_2\text{-}C_3$ π* orbitals, and their HOMOs are their $C_1\text{-}C_2$ π orbitals.[50-52] In MFA and DFA, the electron-donating and -withdrawing characteristics of fluorine operate at the same time but in different ways on the orthogonal π bonds of the molecule. Only the C_1–C_2, substituted double bond can be influenced by the electron-donating properties of the fluorine lone pairs, and, interestingly, empirically

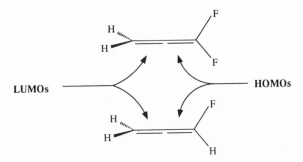

this effect appears to be canceled by the σ orbital polarization (i.e., inductive withdrawing) influence of fluorine on this π bond. This phenomenon has been called the "perfluoro effect" by Brundle et al.[53,54] In contrast, the perfectly aligned, allylic fluorine substituent lowers the π and π* orbital energies of the unsubstituted, $C_2–C_3$ double bond both through inductive withdrawal and through "negative hyperconjugation," that is, by admixture of the π* CF_2 orbital into the π in a bonding fashion.[51]

The dramatically different effect of fluorines on the two π bonds of MFA and DFA along with the thermodynamic factors discussed previously confers on these molecules certain special characteristics. The $C_1–C_2$ π bonds are influenced little by the fluorine substituents, making this π bond, like those of the parent allene, an electron-rich double bond, relatively unreactive in Diels–Alder and 1,3-dipolar cycloadditions, but, owing to the presence of the CF_2 group, probably of enhanced reactivity toward [2 + 2] cycloadditions. In contrast, the $C_2–C_3$ π bonds are strongly stabilized, making this bond, like that of trifluoropropene, very electron deficient and thus highly reactive toward Diels–Alder and 1,3-dipolar cycloadditions, but probably of no special reactivity toward [2 + 2] cycloadditions. Thus it is not surprising that selectivities in attack at one or the other bond have proved to be diagnostic of the nature of the attacking reagent and the mechanism of the reaction. If one had considered the question carefully, there should have been little doubt which double bond of DFA should have been reactive in a Diels–Alder reaction.

3.4. Cycloadditions of 1,1-Difluoroallene

Knoth's results, along with our corroborative efforts, certainly had demonstrated DFA to be an extraordinarily reactive dienophile, and armed with a general understanding of the expected behavior of DFA in Diels–Alder reactions, we embarked upon a detailed study of the regiochemistry of DFA's reactions with unsymmetrical 1,3-dienes, a study which reaped unexpected rewards because it was found that [2 + 2] cycloadditions occurred in competition with the Diels–Alder reaction.[50]

Diels–Alder Reactions

1,3-Dienes: Insight from Regiochemistry. A number of cycloadditions were examined, but for our purposes the results from the reactions of DFA with

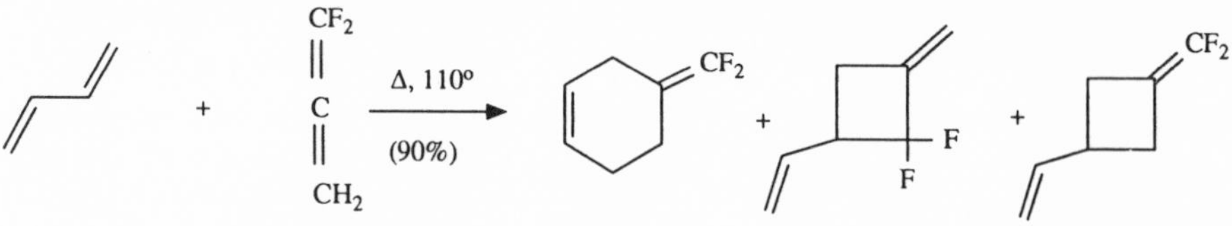

11   **DFA**   **12(63%)**   **13(37%)**   **14(<2%)**

1,3-butadiene and with 2-methyl-1,3-butadiene will suffice. As the reaction with 1,3-butadiene exemplifies, these reactions produced significant amounts of both Diels–Alder and [2 + 2] adducts.

As in the cyclopentadiene reaction, this and other Diels–Alder reactions of DFA were found to be totally regiospecific with respect to the allene, with cycloaddition occurring *only* with the non-fluorine-substituted C_2–C_3 π bond of DFA, in this case to form adduct **12**. Interestingly, the major, competitively formed [2 + 2] adduct (**13**) was found, in contrast, to be that which derived from cycloaddition to the fluorine-substituted double bond. Unlike the [2 + 4] process, the [2 + 2] process was apparently not totally regiospecific since traces (<2%) of the alternate regioisomer **14** could be observed in the ^{19}F NMR of the crude product mixture. [As we will see in later sections, such high regioselectivities as those observed in this [2 + 2] reaction of butadiene and that of isoprene (discussed subsequently) are the exceptions rather than the rule for [2 + 2] cycloadditions of DFA.]

By means of DFA's reactions with unsymmetrically substituted 1,3-dienes such as isoprene (**15**), we were able to examine the regioselectivities of the observed [2 + 4] and [2 + 2] cycloadditions with respect to the diene. As one can see, the observed regioselectivities are quite different for the [2 + 4] and the [2 + 2] cycloadditions, with little selectivity seen for the putative concerted

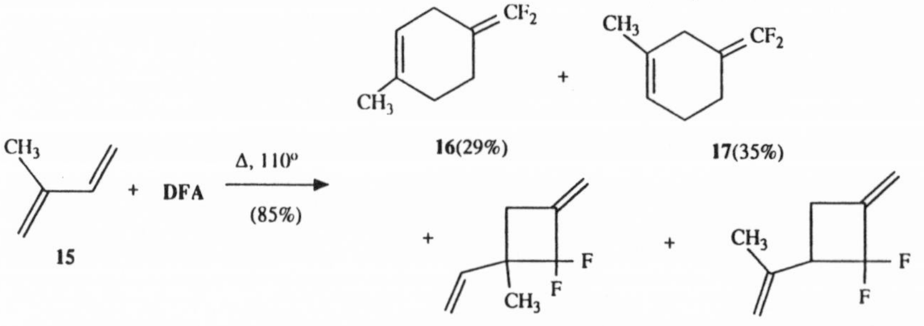

16(29%)   **17(35%)**

15   + DFA   **18(26%)**   **19(10%)**

Diels–Alder reaction, wherein relatively high selectivity is observed for the [2 + 2] reactions. The significant preference for formation of [2 + 2] adduct **18** over adduct **19** is, of course, consistent with the formation of the more stable diradical **20** in preference to diradical **21**.

The observed highly different regioselectivities for these competitive [2 + 2] and Diels–Alder reactions, both with respect to the allene and with respect to the diene, make a common reactive intermediate a very unattractive proposition. The simplest rationale for the results is that the reaction involves a concerted [2 + 4] process in competition with a nonconcerted [2 + 2] process. As mentioned earlier, the [2 + 2] results are completely consistent with the competitive involvement of the two diradical intermediates, **20** and **21**. On the other hand the Diels–Alder results can be nicely rationalized in terms of a concerted process using the frontier molecular orbital theory of cycloadditions.[48,49] Although the C_2–C_3 π^* orbital of DFA is clearly the LUMO, just as importantly calculations indicate that the LUMO coefficients, 0.73 at C_2 and –0.78 at C_3, are nearly identical.[50,51] Thus little selectivity in formation of adducts **16** and **17** would be predicted, as is observed.

These initial cycloaddition studies engendered in us anticipation that the simple regiochemistry of DFA's cycloadditions might well provide unambiguous mechanistic insight, with concerted and nonconcerted reactions being able to be distinguished by mere identification of products. The systematic studies that are described throughout the remainder of this chapter only serve to enhance the credibility of this hypothesis.

The Quest for a Nonconcerted Example. In order to test the viability of this hypothesis, attempts were made to induce DFA to undergo a *nonconcerted* Diels–Alder reaction, in the hope that the yet unobserved [2 + 4] adduct with cyclization having occurred at the CF_2 end of the allene might be detected. With this in mind, the reaction of DFA with 1,2-dimethylenecyclobutane, **22**, was examined on the chance that one might observe an adduct of the type **23**, which would be indicative of a diradical pathway for the [2 + 4] reaction. The particular cisoid geometry of **22** has been demonstrated to impose an inhibition on its potential concerted [2 + 4] cycloadditions because of the unusually large

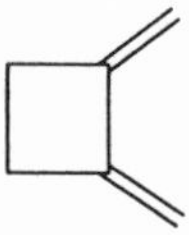

distance (3.35 Å) between the termini of **22** as opposed to those in cyclopentadiene (2.44 Å) or even cisoid butadiene (2.89 Å).[55] While acyclic dienes are
likely to undergo [2 + 2] reactions largely via such extended diradicals as **24**,
which are structurally incapable of forming [2 + 4] adducts, such a rigid cisoid
diene as **22** should form a diradical like **25**, which one might expect to have a
reasonable chance to cyclize to diradically derived [2 + 4] adducts.

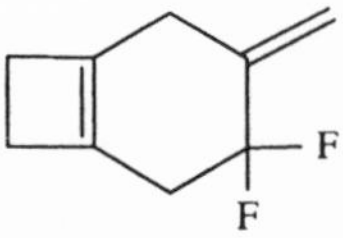

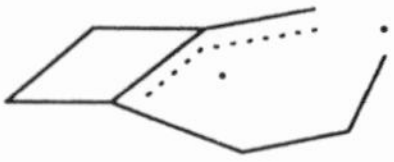

Indeed, as we had hoped, considerable [2 + 2] adduct was formed in the
reaction of DFA with **22**.[56] However, the only observed [2 + 4] adduct was that

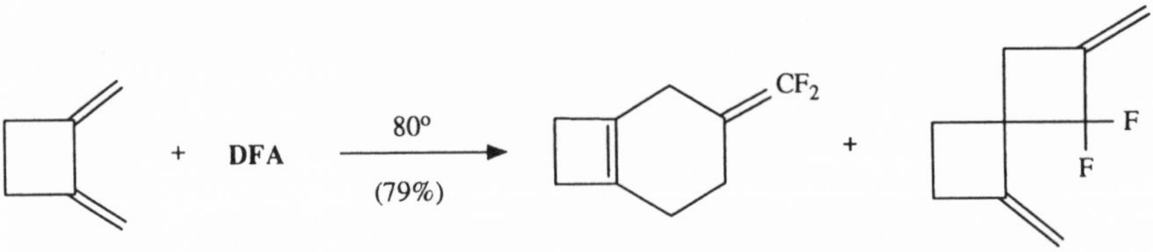

(**26**) which was expected from the concerted reaction.

In looking at the reaction of DFA with the electron-deficient diene, 2,3-
dicyano-1,3-butadiene (**28**), it was expected that the concerted cycloaddition
would be inhibited with concomitant enhancement of all diradical processes.
While the [2 + 4] reaction was indeed significantly diminished, with the [2 + 2]

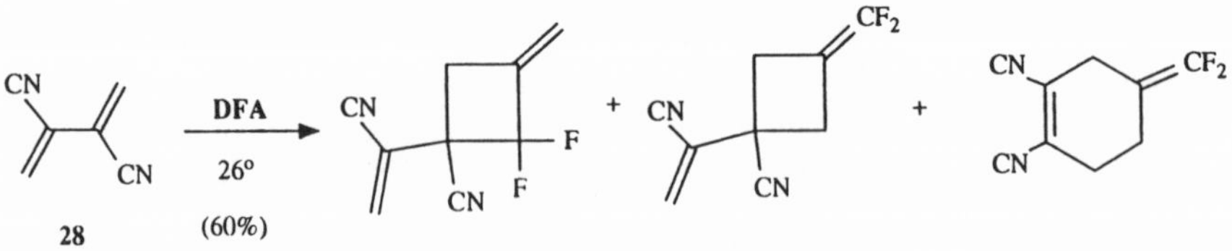

84 : 4 : 12

products dominating, products deriving from the elusive nonconcerted [2 + 4]
cycloaddition process again could not be detected.

In each of these reactions, as was observed before, the major (but in the latter
case not only) [2 + 2] adducts proved to be those with the CF_2 group in the ring,
the position of greater thermodynamic stability.

It is also worth commenting that in reactions of DFA with an assortment of
electron-deficient dienes including **28**, hexachlorocyclopentadiene,

tetrachlorothiophene dioxide, and diphenyl tetrazine, we were never able to induce the DFA to switch roles and react with its C_1–C_2 double bond.[57]

1,3-Dipolar Cycloadditions

Like the Diels–Alder reaction, 1,3-dipolar cycloadditions are $4\pi + 2\pi$ one-step multicenter reactions wherein the 1,3-dipoles play the role of the 4π system, much as the diene does in the Diels–Alder reaction. As such, 1,3-dipolar cycloadditions are allowed by the Woodward–Hoffmann rules to occur by a concerted suprafacial–suprafacial cycloaddition mechanism. The factors that govern the reactivity and regioselectivity of 1,3-dipolar cycloadditions are the same as those that govern the Diels–Alder reaction.[49,58,59] However, for whatever reason, 1,3-dipolar cycloadditions tend to exhibit much more regioselectivity than their Diels–Alder counterparts. This greater regioselectivity and the reasons for it will be discussed throughout this section. The selectivity will be seen to be due in great part to the polar nature of 1,3-dipoles.

Indeed, 1,3-dipoles are species that can *only* be represented by zwitterionic octet resonance structures. When they combine in a cycloaddition with a 2π electron system (the dipolarophile), they form an uncharged five-membered ring.[60,61] The types of 1,3-dipoles we will be discussing, such as nitrones, diazo compounds and nitrileoxides, are all species that react especially well with electron-deficient dipolarophiles. Hence, the important frontier molecular orbital interaction is expected to be HOMO (dipole)–LUMO (dipolarophile), and one would expect such dipoles to be quite reactive with 1,1-difluoroallene.[49]

More Insight from Regiochemistry. DFA was found be highly reactive in 1,3-dipolar cycloadditions, and it underwent facile, high-yield and totally regiospecific cycloaddition to its C_2–C_3 double bond with all 1,3-dipolar reagents that were investigated. Generally, it was also found that a *single* orientation of the 1,3-dipole was also highly, if not exclusively, favored in the cycloaddition, quite unlike our observations for DFA's reactions with 1,3-dienes, which showed little if any orientational selectivity.

In DFA's reactions with nitrones, for example, a *single* regioisomeric product was obtained in all cases.[62,63] The reactions with *N-tert*-butyl-

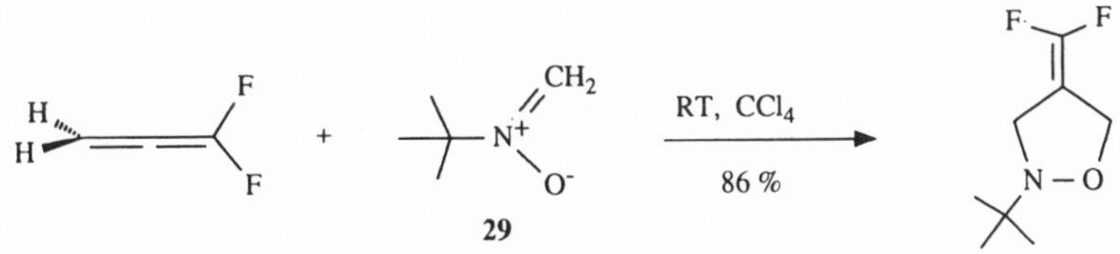

29

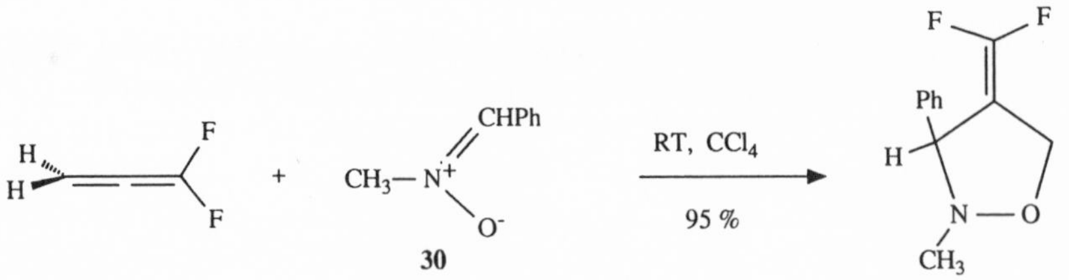

nitrone (**29**) and C-phenyl-N-methylnitrone (**30**) are given as typical examples.

Likewise, DFA's reaction with diazomethane was found to be totally regiospecific.[64]

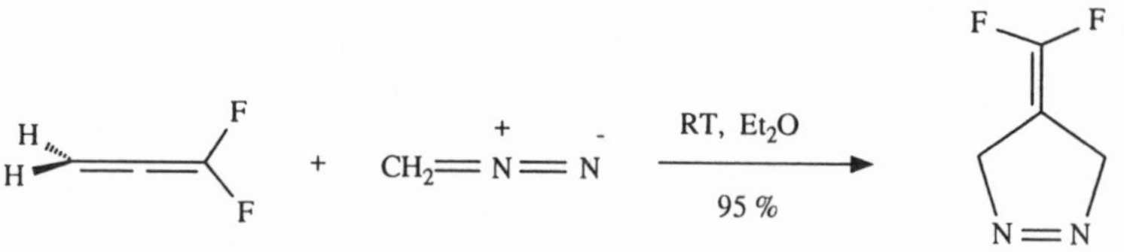

The Factors That Control 1,3-Dipolar Regiochemistry. Why are these reactions so regioselective when Diels–Alder reactions are not? According to a perturbational theory analysis of molecular interactions,[65,66] three forces are potentially operative in the determination of the regiochemistry of bonding in any cycloaddition reaction: (1) electron density on one reactant interacting repulsively with that on the other (nonbonded, steric repulsion); (2) occupied molecular orbitals on one reactant mixing with unoccupied orbitals on the other (frontier molecular orbital interactions); or (3) atoms in one reactant with net positive charge attracting atoms in the other with net negative charge and repelling atoms with net positive charge (electrostatic interactions). The importance of molecular orbital interactions in determining reactivity and regioselectivity in 1,3-dipolar cycloadditions has been discussed in a very important paper by Houk et al.,[67] whereas the potential significance of electrostatic interactions in the determination of regiochemistry of concerted cycloadditions has been discussed in a series of papers by Hehre and co-workers.[68–70]

As was discussed in the section on Diels–Alder reactions, and for much the same reasons that Diels–Alder reactions do not exhibit much regioselectivity, one would not expect frontier molecular orbital interactions to be the source of the high selectivity observed in our nitrone/DFA cycloadditions. The terminal coefficients of diazomethane's HOMO are much less disparate (i.e., 0.775 versus 0.626)[64,71] than those of the dienes (approximately 0.352 versus 0.103 for C_1 and C_4, respectively),[68] which were nonregioselective in their additions to DFA. Thus, the relative lack of regioselectivity in Diels–Alder reactions of DFA with unsymmetrically substituted dienes having significantly different

HOMO coefficients at C_1 and C_4 would seem to speak strongly against frontier molecular orbital effects being determinant in the highly regioselective reactions of DFA with 1,3-dipoles.

If steric factors were to have been important, one would have expected the more sterically significant carbon end of the nitrone (especially when phenyl substituted) to have shied away from the C_2 of the allene, with the consequence that the other regioisomer (not seen) should have been favored.

In the end it appears as if electrostatic interactions are the most attractive rationalization for the regioselectivity observed, not only for this dipole but for the others that we will discuss. Making predictions based upon electrostatic interactions requires simply matching the "nucleophilicity" of the

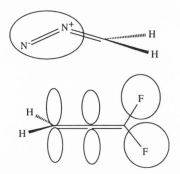

1,3-dipole with the "electrophilicity" of the dipolarophile (i.e., DFA). In this regard, the "negative" ends of the 1,3-dipoles (i.e., the terminal Os of the nitrones and the terminal N of the diazomethane) would be expected to be repulsive to the negative region of DFA (i.e., the fluorine substituted end), with the result that the observed preferred orientation of addition should be and is where these negative ends of the 1,3-dipoles are oriented toward the CH_2 end of DFA.

Further 1,3-Dipolar Results and Discussion. In contrast to the above results with the parent diazomethane, the cycloadditions of *substituted* diazomethanes (**31**) with DFA were *not* regiospecific, but led to two adducts,

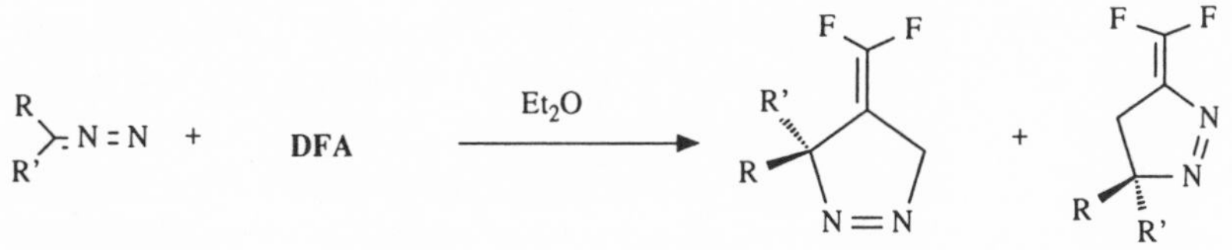

R
>=N=N + **DFA** $\xrightarrow{\text{Et}_2\text{O}}$
R'

31-a R = R' = Me

b R = R' = fl =

c R = R' = Ph

32-a,b,c   +   **33-a,b,c**

Table 1.   Regiochemistry of Diazoalkane Cycloadditions to DFA

Diazoalkane	Conditions	Relative Yield of Regioisomers		Total Yield (%)
		32	33	
$CH_2=N_2$	RT, 5 min	>99	—	95
31a	0°, 5 min	61	39	99
31b	RT, 4 h	28	72	99
31c	28°, 5 h	14	86	95

32 and **33**, these deriving from the two possible orientations of the nitrone with respect to the C_2–C_3 π bond of DFA (Table 1). What can be seen is that as the substituents on the diazo compound become increasingly bulky, adduct **33** becomes more favored as the product.

These regiochemical results can be rationalized in terms of two competing effects: one, electrostatic interactions, which, as discussed earlier, would appear

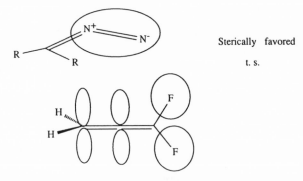

Sterically   favored

t. s.

intrinsically to favor formation of isomer **32,** and a steric effect that clearly favors adduct **33.** Thus, as the substituents increase in size, steric effects are seen to intervene and compete effectively in determining the regiochemistry of the reaction. Such a reversal of regiochemistry had been observed before in the addition of diazo compounds to thiete 1,1-dioxide.[71]

Naively we expected that the relative importance of electrostatic interactions in determining regiochemistry of cycloaddition might be altered by modifying the polarity of the solvent in which the reactions were carried out. Much to our surprise we found that the ratio of products (i.e., **32** : **33**) from the reaction of diphenyldiazomethane with DFA was identical (14.4 : 85.6) regardless of whether the reaction was carried out in DMSO (dielectric constant = 46.7) or in hexane (dielectric constant = 1.9).[64] This led to the following question: Why does one not observe an effect of solvent polarity on the influence of electrostatic interactions?

Nitrile oxides are a classic type of 1,3-dipole. DFA's reactions with two nitrile oxides, phenyl nitrile oxide and mesityl nitrile oxide, were investigated. Interestingly, *both* regioisomeric adducts were formed in the reaction with phenyl, but not with mesityl nitrile oxide. Obviously this variation in behavior cannot

$$R \equiv\!\!\equiv N^+\!-O^- \quad + \quad \textbf{DFA} \quad \xrightarrow{\text{CCl}_4,\ \text{RT}}$$

			35	**36**
			56%	44%
34-a	R = Phenyl	1 h(75%)		
34-b	R = Mesityl	12 h (94%)	>99%	

be due to a differential steric effect, since the more sterically hindered transition state should be that one leading to adduct **35**, which is actually the one preferentially formed in the mesityl case. Again, electrostatic interactions can be used to rationalize the results. First, why should *any* of the regioisomer **36** have been formed in the phenyl case? Our guess is that, owing to the linear nature of a nitrile oxide, the phenyl substituent would be sticking right in the face of the CF$_2$ group of DFA in the transition state leading to the "normal" product **35**, and that it is the resultant repulsion of the phenyl π electrons that in the end makes competitive the alternate transition state that would lead to adduct **36**. Why then should the mesityl nitrile oxide not give rise, at least qualitatively, to the same effect? Again, our guess is that the electron-donating ability of the three methyl substituents of the mesityl group must enhance the net π-donating ability of the phenyl group so that the terminal oxygen of the nitrile oxide has more net negative charge and thus greater repulsion for the CF$_2$ group, with the result that adduct **35** should be more preferred in the mesityl case.

Conclusions. In summing up the results of the reactions of 1,3-dipoles with 1,1-difluoroallene, most important is the observation that the overall regioselectivities exhibited in these reactions appear to be totally consistent with a concerted, pericyclic mechanism. Although most workers in the field would be comfortable with that conclusion, Firestone has over the years effectively championed consideration of a stepwise mechanism.[72,73] Unfortunately for the Firestone hypothesis, we know from our previously described competitive [2 + 4] and [2 + 2] cycloaddition reactions of DFA with dienes that the regiospecific reaction of the 1,3-dipoles to the C$_2$–C$_3$ double bond of DFA is diagnostic of a concerted reaction. Referring also to the diene study, we know that if diradical intermediates were involved, as they most certainly were in the competitive [2 + 2] reactions of DFA, one would have expected predominant formation of adducts in which the CF_2 group would have been incorporated into the five-membered ring. The improbability that a nonconcerted reaction can actually be involved is emphasized by the fact that no traces of such adducts could be detected in any of the 1,3-dipolar cycloadditions that were studied.

Therefore, it would appear that, while there may still be discussion as to whether electrostatic interactions are primarily responsible for the orientational preferences of the dipoles, there can be little doubt that these reactions are pericyclic and not multistep processes.

[2 + 2] Cycloadditions: Reactivity and Selectivity

In addition to those insightful [2 + 2] reactions between DFA and 1,3-dienes, which were found to be taking place in competition with Diels–Alder reactions, DFA was observed to be considerably more reactive than allene itself in reactions with normal [2 + 2]-reactive addends, such as

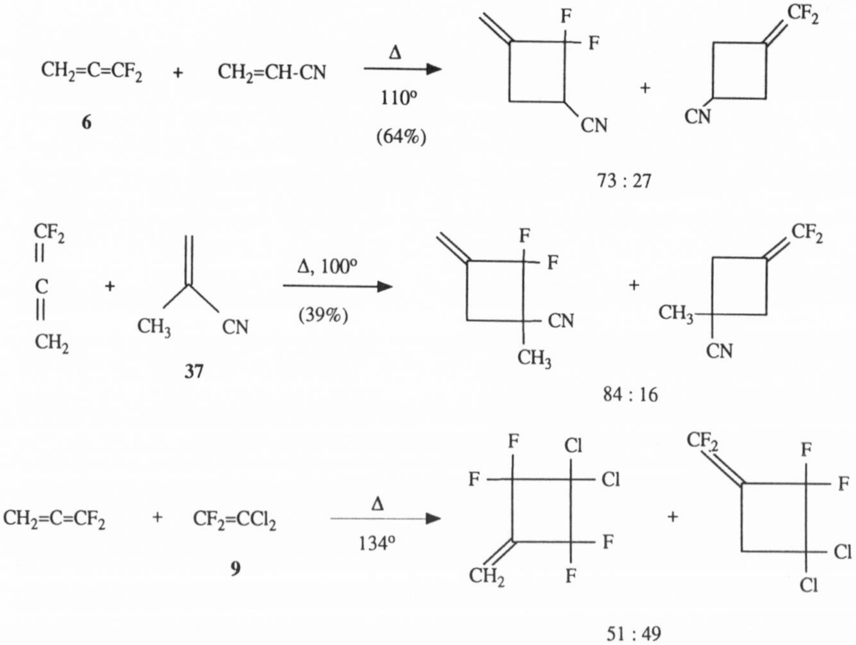

$$CH_2=C=CF_2 \quad + \quad CH_2=CH\text{-}CN \xrightarrow[110°]{\Delta}$$

6   (64%)

73 : 27

$$\begin{array}{c} CF_2 \\ \| \\ C \\ \| \\ CH_2 \end{array} + \xrightarrow[(39\%)]{\Delta,\ 100°}$$

37

84 : 16

$$CH_2=C=CF_2 \quad + \quad CF_2=CCl_2 \xrightarrow[134°]{\Delta}$$

9

51 : 49

acrylonitrile, methacrylonitrile, α-acetoxyacrylonitrile, and 1,1-dichloro-2,2-difluoroethylene.[56,57,74] For example, whereas allene required heating at a temperature of 210°C to accomplish its reaction with acrylonitrile,[32] DFA required only a temperature of 110°C, and their reactions with 1,1-dichloro-2,2-difluoroethylene required 170° and 134°C, respectively.

Since in all of the above reactions variable degrees of dimerization and oligimerization of DFA were observed, the dimerization reaction was examined in detail. After four days at room temperature, half of the DFA had reacted, and only three volatile products, **38**, **39**, and **41**, were obtained in addition to polymeric material. The observed products raise some very interesting ques-

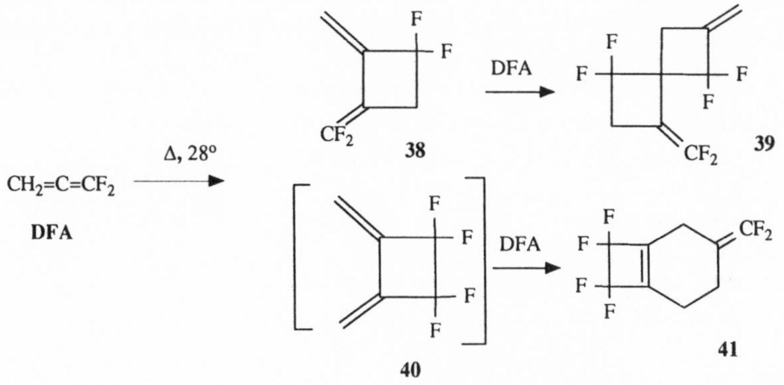

38:39:41 = 33:17:49

tions, which have not yet been answered. From their structures it would appear that dimer **38** undergoes *only* a [2 + 2] reaction, and that that reaction occurs *only* at the CH$_2$ end of the diene to form trimer **39**. On the other hand trimer **41** would appear to have derived from unobserved dimer **40**, which itself apparently *only* undergoes a Diels–Alder reaction. The specificity of these reactions has yet to be explained.

It can be seen that all of these [2 + 2] cycloadditions of DFA proceed basically as one would have expected for reactions involving diradical intermediates— that is, mixtures of regioisomeric products are obtained with the ratio of products generally being consistent with (1) the relative stabilities of the products and (2) the relative stabilities of the potential diradical intermediates. In reference to the latter factor, although the number of examples is certainly limited, there seems to be a rough correlation between the stabilities of the diradical intermediates and the product ratios. This correlation may be explained by applying the Hammond postulate to closure of the diradical intermediate. The longer-lived, more stable the diradical, the later the transition state and the more the transition state should resemble the products. Thus with the demonstrated thermodynamic preference for geminal fluorines to be located on an sp^3 carbon versus an sp^2 carbon, such a later transition state should lead to increasing closure of CF$_2$ to put the fluorines on the cyclobutane ring as the diradical intermediate becomes more stable.

Although the results presented thus far are indeed consistent with a stepwise mechanism for the [2 + 2] cycloadditions of DFA, that does not really clarify the details of the process of getting from starting materials to products. In fact, the mechanisms of [2 + 2] cycloadditions have in detail proved quite inscrutable. Little theoretical effort has been directed at them,[75] and not much more is known about the factors that drive such reactions than the general rules espoused in Roberts and Sharts's classic review of all known [2 + 2] reactions in 1962.[76] [2 + 2] cycloadditions do not appear to be driven by HOMO–LUMO (donor–acceptor) interactions since many if not most of the best [2 + 2]s are dimerizations. In general, classic factors seem to promote [2 + 2] reactivity, that is, relief of strain and/or the presence of radical-stabilizing or olefin-destabilizing substituents.

As mentioned earlier, allenes are especially advantageous addends for the study of [2 + 2] cycloaddition mechanisms. Although free-radical additions to allenes occur at both terminal and central carbons,[77–79] [2 + 2] reactions of allenes proceed virtually exclusively via initial central carbon attack.[80] The contrast in regiochemistry for these superficially similar processes can be understood in terms of the differences in energetics for the two processes. The addition of radicals to allene is generally an exothermic process with a very low activation energy and with little observed temperature dependence on regiochemistry. Steric effects have been shown to play a dominant role in determining the regiochemistry of such reactions.[77–79] In contrast, the initial step in a [2 + 2] cycloaddition of allenes is endothermic with a significant activation energy. This, combined with the regiospecificity observed in initial bond formation, leads one to conclude that the transition state for this first step likely has undergone significant rotation of the terminal methylene of the reactive double bond, giving the transition state considerable stabilizing allyl radical character.

From our studies of the [2 + 2] reactions of DFA one can be reasonably sure that the initially reactive π bond of DFA is the fluorine-substituted one. This assumption derives from the observation that fluorine-substituted olefins are much more reactive in [2 + 2] cycloadditions than are hydrogen-substituted ones.[76] Especially significant is the comparison of the [2 + 2] reactivities of methylenecyclopropanes **42** and **43**.[81] These species are "homoallenes" and should be considered excellent models of the fluorine-substituted and the hydrogen-substituted π bonds, respectively, of DFA. As one can see, the isomer with the fluorine-substituted double bond (**42**) is considerably more reactive. Thus, in the [2 + 2] cycloadditions of DFA, we will generally assume that initial carbon–carbon bond formation is to the C_1–C_2 π bond. Of course, once the allyl radical is formed, the two ends of the allyl system become indistinguishable with respect to which bond had undergone reaction.

Moreover, once the intermediate is formed, only rarely has it been observed to revert to starting materials,[82] unlike the usual [2 + 2] diradical intermediates, because of the necessity that this same allyl radical stabilization be lost prior to such cleavage. We indeed have observed *no* reversibility of this initial bond formation under conditions of cycloaddition in any of our studies of the [2 + 2] reactions of allene or fluorine-substituted allenes.

The formation of such an allylethyl diradical intermediate as **44** (shown for the reaction of DFA with acrylonitrile, which has the potential for ring closure at *two* sites) gives rise to yet another unique aspect of allene [2 + 2] cycloadditions: Unlike the case with virtually all other alkenes,

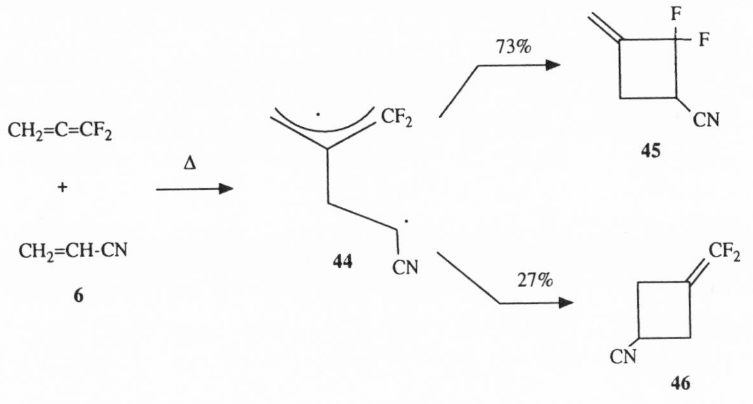

the regiochemistry of the [2+2] cycloaddition of an unsymmetrical allene is *not* totally determined at the point of initial bond formation. For example, intermediate **44** can cyclize to form either adduct **45** or **46**, with **45** in this case being significantly predominant. The presence of such a branch in the mechanism provides one with enhanced opportunity for mechanistic probing, and for our purposes the DFA/styrene reaction system was chosen as the one with the most potential for kinetic and stereochemical analysis.

The DFA/Styrene Reaction System

In order to obtain more direct evidence for the intermediacy of diradicals in allene [2 + 2] cycloadditions, as well as to gain insight into the kinetic behavior of such diradicals, a reactivity and stereochemical examination of the

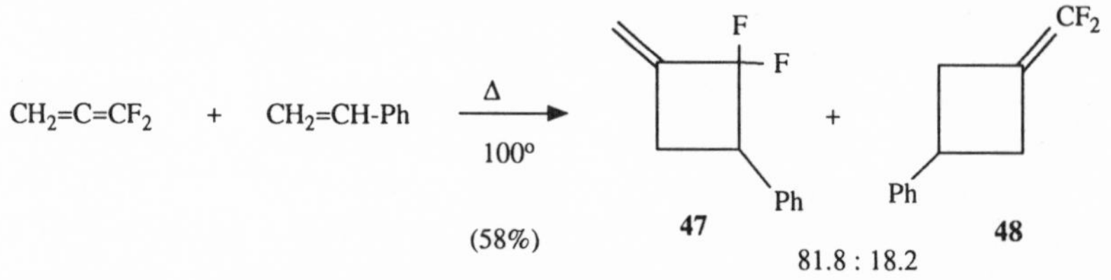

$$CH_2=C=CF_2 \quad + \quad CH_2=CH-Ph \quad \xrightarrow[100°]{\Delta}$$

(58%)

47

81.8 : 18.2

48

DFA/styrene reaction was carried out. Styrene underwent cycloaddition with DFA cleanly at temperatures as low as 60°C to yield significant amounts of both regioisomeric adducts, **47** and **48**.[83] The lack of a higher yield was due to the observed inevitable, competitive, but unobtrusive oligomerization of DFA.

A slight temperature dependence in the ratio of **47** to **48** was observed, with the greatest selectivity, not surprisingly, being observed at the lowest temperature. The ratio diminished from 5.6 : 1 to 4.5 : 1 as the temperature was raised from 60° to 100°C.

Reactivity Study. In a cursory Hammett-type study of reactivity, the reactions of DFA with *p*-methoxy- and *p*-nitrostyrene were examined with regard to relative reactivity and ratios of products. These substituted styrenes proved to be both more selective and more reactive than styrene, with the product ratios for *p*-methoxy- and *p*-nitrostyrene being 83.1 : 16.9 and 83.3 : 16.7, and their reactivities relative to styrene, 1.3 and 2.1, respectively. Such results are totally consistent with the intervention of diradicals in the reaction, inasmuch as both the nitro and the methoxy substituents have been shown to enhance the rate of radical formation.[84] Note also that the slightly greater regioselectivity that is observed for the nitro- and methoxystyrene cycloadditions is consistent with *more stable* diradical intermediates being involved.

Stereochemical Study. Our examination of the stereochemistry of the reaction of (Z)-β-deuteriostyrene with DFA yielded results that were even more exciting than we had expected. Indeed our observations, which are given in

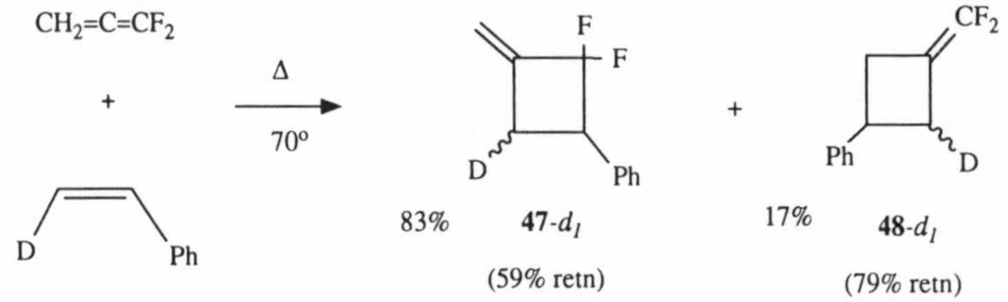

$$CH_2=C=CF_2$$

+

D————Ph

$$\xrightarrow[70°]{\Delta}$$

83%   **47**-*d₁*

(59% retn)

17%   **48**-*d₁*

(79% retn)

Table 2. Reaction of DFA with (Z)-β-Deuteriostyrene:
Stereochemistry and Product Ratios

		Stereochemical Ratios	
Temperature (°C)	Product Ratio (47 : 48)	47-Z : 47-E	48-Z : 48-E
100	81 : 19	57 : 43	79 : 21
80	82 : 18	58 : 42	79 : 21
70	83 : 17	59 : 41	79 : 21

Table 2, provided us with the first clue that this reaction system had the potential to provide us with new and important mechanistic insight into the complexities of [2 + 2] cycloadditions.

Note that the reaction occurs with a significant but quantitatively very different loss of stereochemistry being observed in each of the products, a result that, while totally consistent with a nonconcerted mechanism, is totally *incon*sistent with there being a *single* common intermediate for the two products. As it turns out, it is the presence of the extra regiochemically significant branch in the DFA [2 + 2] mechanism that has allowed one to scrutinize aspects of these diradical mechanisms that have heretofore proved to be inscrutable.

In this reaction, the two products, **47** and **48**, were obtained with different degrees of retention of stereochemistry. Such results are inconsistent with a single diradical intermediate being involved, and can only be rationalized by a mechanism involving two kinetically distinct intermediates. Scheme 1 depicts a possible mechanism. In this mechanism, it is assumed that these two diradical "intermediates" were actually two conformationally related *systems* of diradicals, A-A′ and B-B′, which exhibit different kinetic behavior. (Actually, whereas this scheme depicts DFA being attacked from below, an enantiomeric scheme could be drawn wherein the DFA would be attacked from above.) In the scheme, A and A′ are depicted as initially formed diradicals with structures consistent with minimal conformational change. Such diradicals are projected to be highly reactive, and as such should exhibit relatively low regiochemical selectivity in cyclization and should, upon cyclization, lead to products with high if not total retention of the *cis* configuration of the deuterium substituent relative to the phenyl substituent. Competitive with such cyclization would be a highly favorable conformational conversion to a more stable pair of intermediates. Intermediates B and B′, in contrast, are depicted as having structures within which the phenyl substituent has rotated so as to minimize nonbonded interactions, thus creating diradical intermediates that should be thermodynamically more stable and have less kinetic reactivity. Whereas B and B′ would probably not be formed in equal amounts, they should cyclize with

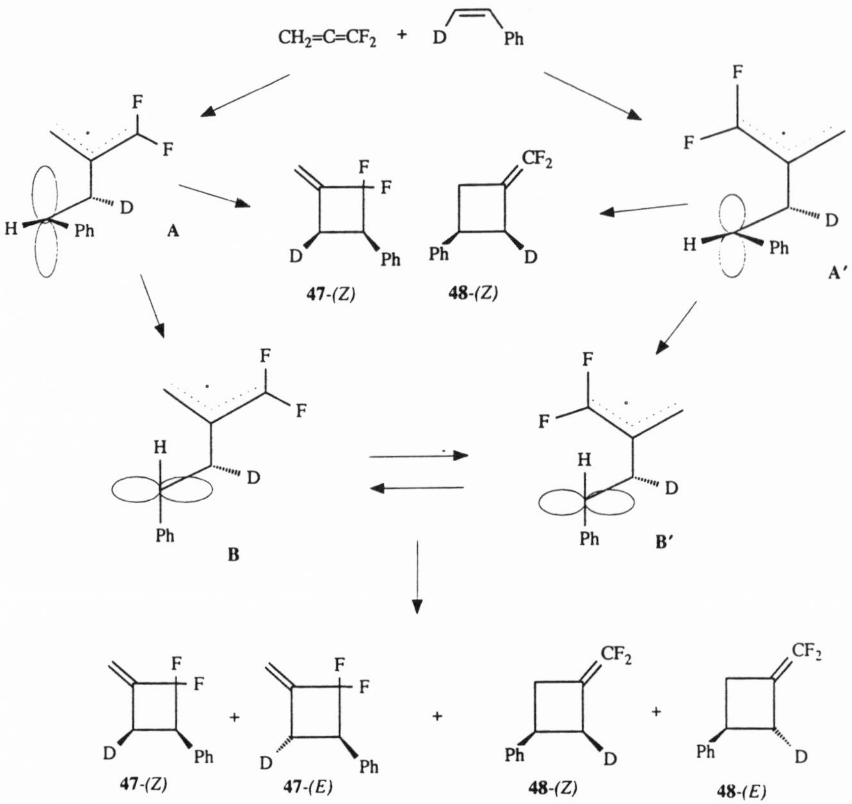

Scheme 1

virtually identical regiochemistries (but ones clearly different from the regiochemistry exhibited by A and A′), with the result that from this mixture of B and B′ products **47** and **48** should be formed in different amounts (i.e., **47** observed to be favored), but with identical stereochemistries. Therefore if the products had derived only from intermediates A and A′, or only from intermediates B and B′, they should have been formed with identical stereochemistries. The contrary observed results are, however, consistent with the mechanistic scenario depicted in Scheme 1, wherein both of the diradical intermediate systems, A, A′ and B, B′, each expected to exhibit different regiochemical and stereochemical behavior, are involved in the formation of products.

In our mechanistic scheme we have imposed certain reasonable but arbitrary kinetic characteristics upon these kinetically distinct systems of diradicals in order to explain the results. Diradicals A and A' are characterized as being highly reactive species that cyclize with high stereoselectivity to products **47** and **48**, but that mostly are projected as converting to a more stable, less reactive system of diradicals, B, B', which themselves cyclize to **47** and **48**, with less stereoselectivity. In fact, if one simply assumes total retention of stereochemistry for cyclization from A and total loss of stereochemistry for cyclization from B, one can calculate a mechanistic scenario within which the observed results would be obtained if 25% of the products were formed from intermediates A and A' (15% of product **47** and 10% of product **48**), while 75% of the products came from diradicals B, B' (68% of product **47** and 7% of product **48**). One can see that, consistent with the mechanistic model, the more reactive intermediates (A, A') are seen to cyclize less regioselectively (i.e., 3 : 2), than the less reactive intermediates (B, B') (9.7 : 1).

One could have imposed less extreme arbitrary stereochemical requirements upon the intermediates, in which case one would have obtained qualitatively similar but quantitatively different results. The *qualitative* nature of Scheme 1 is required, however, if one is to explain the results satisfactorily. The diradicals proposed in Scheme 1 are, as stated earlier, depicted simply as conformational isomers. There is no direct evidence for the actual structures; indeed the stereochemical results merely require two pathways to products **47** and **48**, a minor pathway that is nonregioselective and stereospecific and a major pathway that is more regioselective but stereorandom. However, it should be remembered that the specific mechanism depicted and discussed above is one that is consistent with everything that is presently known about allene [2 + 2] cycloaddition reactions.

Pasto and Yang, in a related stereochemical study of the reactions of 1,1-

dimethylallene with diethyl fumarate and diethyl maleate, reached similar conclusions in a much more complex reaction system.[82]

Effect of Styrene α-Substituents. In order to probe further the details of the previously proposed mechanism, with the aim of more fully elucidating those factors that should have an impact upon the regiochemical and stereochemical outcome of the DFA/styrene reaction, we examined the systematic perturbation

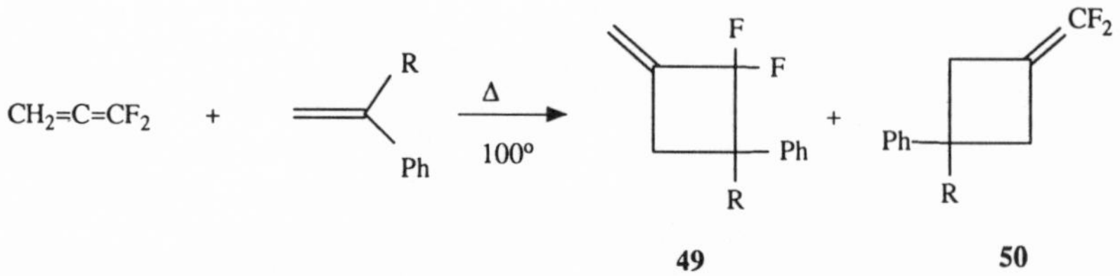

$$CH_2=C=CF_2 \quad + \quad \text{(structure with R, Ph)} \quad \xrightarrow[100°]{\Delta} \quad \text{(49)} \quad + \quad \text{(50)}$$

of the system through substitution of the styrene component at the α position.[85] It was considered probable that an increase in radical stabilizing ability would decrease the reactivity of both proposed intermediates, A and B, with the specific result that A would be longer-lived and have a higher probability of converting to the more stable and less regioselective B, much as was the case when radical stabilizing substituents (NO_2 and OCH_3) were placed on the styrene component.

Therefore it was expected that substituents such as chlorine or phenyl should lead to greater regioselectivity in the reaction. This proved to be the case. On the other hand, what should one have expected to happen if one increased the steric bulk of the α substituent without changing the radical stabilization? In fact, as Table 3 indicates, the modification of R from H to methyl, ethyl, isobutyl, and finally isopropyl gave rise to, at first, a gradual diminution of regioselectivity, but with the bulky isopropyl the ratio of products actually dropped past the "magic" 50 : 50 mark, which would be indicative of complete nonregioselectivity, and **50** was observed to be the major product. The mechanistic model depicted in Scheme 1, being too simplistic to handle such results, needed some fine tuning to deal with situations in which *steric effects* can play a role. Within our model mechanism, we have included only selected conformational equilibria: those that we believe to be of significance in determining the kinetic behavior of the "intermediate" with regard to the ratio of

Table 3.   Regiochemistry for the [2 + 2] Cycloadditions of DFA with α-Substituted Styrenes[85]

		Regiochemistry (%)	
R	*Isolated Yield (%)*	*49*	*50*
H	35	81.1	18.9
Me	37	79.6	20.4
Et	28	70.6	29.4
Isobutyl	22	69.7	30.3
Isopropyl	15	31.3	68.7
F	26	79.6	20.4
Cl	23	85.5	14.5
Ph	18	84.4	15.6

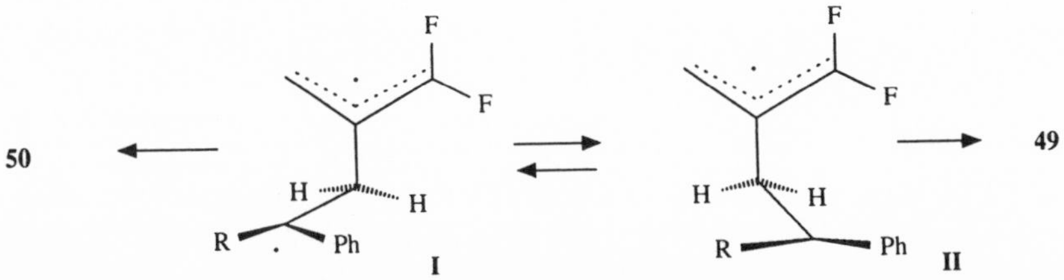

products. However, as the bulk of the ethyl end of the diradical becomes a factor, conformational equilibria such as that depicted here can begin to play an important role in determining the ratio of products. As R becomes increasingly bulky, the less sterically crowded conformational intermediate (**I**) would be expected to become increasingly favored. In all likelihood this less sterically crowded intermediate would favor formation of the less sterically crowded product (**50**). (In fact, with geminal bulky substituents at the 3 position of methylenecyclobutane product **49**, it may be that product **50** now has become thermodynamically favored over product **49**.)

Fluoro versus Methyl Substituent. In a final comment on this investigation, it is interesting to note that α-fluorostyrene and α-methylstyrene *give exactly the same ratio of products* in this reaction. This is now the third example of a diradical system wherein a fluoro substituent and a methyl substituent were

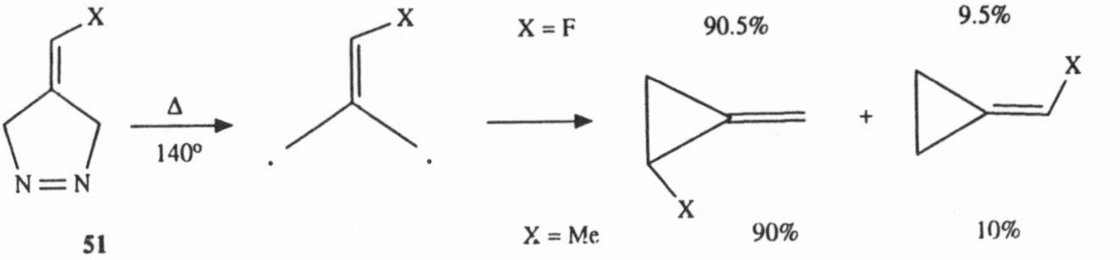

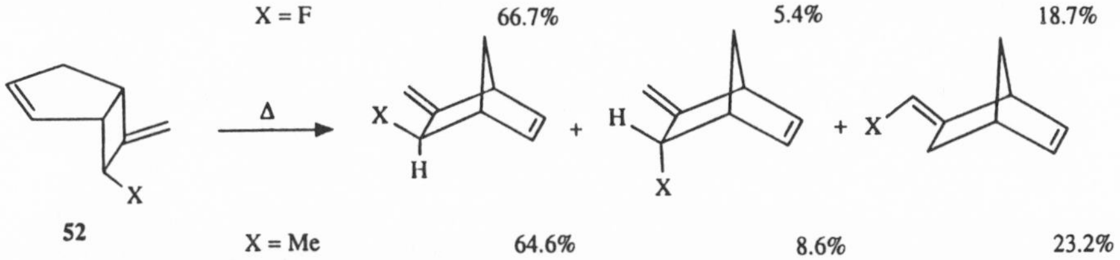

observed to exert *virtually identical* kinetic influence upon the kinetic fate of the diradical. The other examples were very different reactions from the present one, i.e., the deazetations of 4-methylenepyrazoline, **51**,[86-88] and the thermal isomerizations of the epimeric 7-fluoro-6-methylenebicyclo[3.2.0]hept-2-enes, **52**.[89-91]

The Attempt to Find the Perfect Mechanistic Probe. In attempting to demonstrate unambiguously the mechanism of a particular reaction, one must be careful that the probe used to study the mechanism does not itself in fact change the mechanism, with the result that insight into the mechanism of the designated reaction cannot be forthcoming. With this in mind, what kind of probe can one use that will not perturb the mechanism? Modification of the structure of the reactants, by its nature, always has the potential to give rise to mechanistic change, owing either to steric or electronic effects. For example, in the previously mentioned study in which α substituents were attached to the styrene addend, there is every reason to believe that whatever conformational equilibria existed in the parent system would certainly be modified significantly owing to the presence of the substituents. Steric effects, which were utilized to rationalize the results of that study, most often (as they did in that case) prove to be a nemesis to the physical organic chemist in his or her attempts to interpret unambiguously the effect of a particular substituent on the outcome of a reaction. Deuterium "substituents," which do not perturb the potential energy surface of a reaction and therefore cannot give rise to a mechanistic modification, are the perfect substituents to use in a mechanistic study, but as we saw in the first section of this chapter, there can be definite limitations in one's ability to use isotope effects as the sole probe in definitive mechanistic determinations. One needs other equally subtle mechanistic probes that will not perturb the energy surface of the reaction being examined.

High-Pressure Studies. The effect of very high pressure upon the regiochemical and stereochemical outcome of the DFA/styrene [2 + 2] cycloaddition was considered to be exactly the kind of subtle but potent, non-mechanism-modifying probe that might provide the unique insight needed to corroborate the general mechanistic scheme that we had postulated based upon our stereochemical study.

Given $d \ln k/dP = -\Delta V^*/RT$,[92] it was predicted—in view of the expected negative ΔV^* for the cyclization of diradicals A and A' of Scheme 1 as compared to little expected ΔV^* for any rotational conversion of A to B—that under increasingly high pressure the reaction should be induced to derive increasingly from diradicals A and A'. Thus product formation should become less regioselective and more stereoselective.

In our study we examined both the regiochemical and the stereochemical outcomes of the DFA/styrene reaction as a function of pressure, varying the pressure from 1.8 to 13 kbar (a kilobar is equal to 1000 atmospheres of pressure, or 14,700 lb/ft^2). The results are given in Table 4.[93,94] It can be seen that there is a very noticeable effect of pressure upon both the regiochemistry and the stereochemistry of product formation, with the regioselectivity of the reaction diminishing significantly as one increases the pressure, while the stereoselectivity increases for both products. Indeed, at 13 kbar the minor product, **48**, exhibits 95% retention of the *cis* configuration. As indicated in Table 5, again assuming the proposed quantitative mechanistic model, such a degree of

Table 4. Pressure dependence of Product Ratios for the DFA/(Z)-β-Deuteriostyrene Reaction

P (kbar)	47	Z : E Ratio	48	Z : E Ratio
1.8	86.1	66.3 : 33.7	13.9	88.0 : 12.0
4.1	84.1	71.6 : 28.4	15.9	89.9 : 10.1
5.9	82.1	73.2 : 26.8	17.9	91.2 : 8.8
8.0	77.7	74.5 : 25.5	22.3	92.3 : 7.7
11.0	70.6	83.2 : 16.8	29.4	94.6 : 5.4
13.0	69.6	85.5 : 14.5	30.4	95.2 : 4.8

stereochemical retention translates into an increase to about 77% of product formation deriving from intermediate A (as opposed to 25% at ambient pressure).[83] This comprises a very significant perturbation upon the outcome of this reaction, and it speaks very clearly in terms of the mechanistic insight it provides.

The results indicate that the imposition of high pressure upon the cycloaddition reaction gives rise to a pathway that is both less regioselective and more stereoselective. This is consistent with our hypothesis of a highly reactive initially formed intermediate that cyclizes nonregioselectively and highly stereoselectively via a process which at ambient pressure is unfavorably competitive with conformational conversion to a less highly reactive diradical intermediate. With cyclization of A having a negative volume of activation, and its conformational conversion expected to have little ΔV, the imposition of high pressure would be expected to lead to results such as those that were observed. Recognizing that $(d \ln k_A/k_B)/dP = -\Delta\Delta V^{\ddagger}_{A-B}/RT$,[92] one can calculate the apparent difference in activation volumes for the two competitive paths to be -3.6 cm^3/mol.

Table 5. Calculated Fractions of Products Deriving from Intermediates A and B (Assuming Mechanistic Model)

P (kbar)	Percent Total Product		Percent Product 47		Percent Product 48	
	From A	From B	From A	From B	From A	From B
1.8	38.7	61.3	28.1	58.0	10.6	3.3
4.1	49.1	50.9	36.4	47.7	12.7	3.2
5.9	52.8	47.2	38.0	44.1	15.0	3.2
8.0	56.9	43.1	38.1	39.6	18.9	3.4
11.0	73.1	26.9	46.8	23.8	26.3	3.2
13.0	76.9	23.1	49.5	20.2	27.4	2.9

In terms of the details of our hypothesized mechanism, as mentioned earlier, one cannot be certain as to the *structures* of intermediates A and B. The present work does, however, give clear insight into the dynamic nature of these intermediates. Again, while our choice of extreme stereochemical behavior for the intermediates (i.e., total stereospecificity for cyclization from A and total lack of stereoselectivity for cyclization from B) is quantitatively arbitrary, it indeed cannot be far from the fact in view of our observation that the minor product, **48**, exhibits such high stereoselectivity (95%) for its formation at 13 kbar.

The Potential Importance of Viscosity.   Some time ago, Firestone proposed that observed effects on rate that were attributed to negative volumes of activation could just as well be rationalized as deriving from viscosity effects.[95] Although we have not carried out definitive experiments on the effects of viscosity in this reaction, one experiment made it quite clear that the influence of viscosity can play an important role in the effects that we have observed. All experiments reported in Table 4 were carried out without solvent. In an earlier report in which we determined the effect of pressure on the regiochemistry of the DFA/styrene reaction, pentane was used as a solvent, and when the data from both studies were plotted, regiochemistry versus pressure, the two sets of data led to parallel straight lines, a result that was consistent with the former study having been carried out in a less viscous medium. To confirm that an increase in viscosity could indeed give rise to results qualitatively the same as those deriving from an increase in pressure, an additional experiment was carried out at the end of the latter study, using non-deuterium-labeled styrene and pentane as solvent (pentane : styrene = 13 : 1). Indeed, the reaction, which was carried out at 4.1 kbar, was significantly more regioselective, giving a product ratio of 6.25 (86.2% **47**) versus the 5.3 (84.1% **47**) observed in the solventless experiment.   The only logical explanation for the difference in results with and without solvent is the effect of diminished viscosity in the experiments run in pentane. Thus the viscosity of the medium does apparently exert a significant effect upon the regiochemical, and probably would also affect the stereochemical, outcome of this reaction.

The relative importance of viscosity effects versus the effect of differences in volume of activation could not be determined from this study. However, for the purpose of mechanistic interpretation, this question is immaterial. The conclusions that derive from the results are the same regardless of the cause.

Conclusions.   Our multifaceted study of the mechanism of the reaction between DFA and styrene comprises perhaps the most comprehensive and definitive experimental examination of a [2 + 2] cycloaddition extant. Taken in their entirety, the results clearly support the general description of the mechanism that is depicted in Scheme 1. If anything, this mechanism is oversimplified because it does not nearly indicate the potential complexity of conformational equilibria that could be involved in this reaction and others similar to it, equilibria that could, under the right circumstances, have an impact upon the outcome of such reactions. It remains for us to devise still more subtle (and sometimes not so subtle) probes to define further the details of [2 + 2] reaction mechanisms.

3.5. Cycloadditions of Fluoroallene

As was discussed earlier, fluoroallene should be expected to have much the same overall reactivity characteristics as 1,1-difluoroallene. As was the case for DFA, the HOMO of fluoroallene (MFA) is its C_1–C_2 π orbital, and the LUMO of MFA is its C_2–C_3 π^* orbital. According to semi-empirical, i.e., INDO,[63] and ab initio[52] calculations, MFA should be intermediate in dienophilic and dipolarophilic reactivity between allene and DFA, the LUMO of MFA being significantly stabilized (lower in energy) relative to allene, but not as stabilized as the LUMO of DFA. One would therefore predict that, in its Diels–Alder and 1,3-dipolar cycloadditions, MFA should undergo reaction regiospecifically with its C_2–C_3 double bond.

Indeed, early studies of the Diels–Alder reactions of MFA substantiated our expectations as to its reactivity and regioselectivity. MFA was observed to react with cyclopentadiene at 0°C over a period of four days (DFA reacted instantly

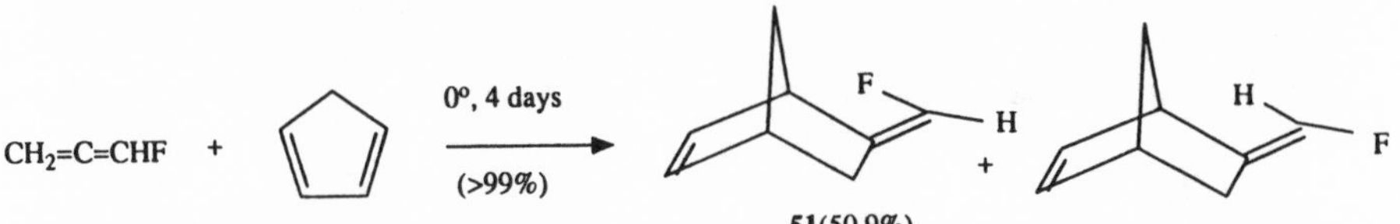

at 0°) to produce virtually quantitatively an almost 50 : 50 mixture of the two possible adducts, **51** and **52**, that one could obtain from addition to the its C_2–C_3 π bond.

The Element of π-Faciality

The most interesting aspect about the cycloaddition reactions of MFA, however, has to do with the fact that there is a *stereochemical* element to these reactions, since products that derive from C_2–C_3 addition will necessarily have exhibited either a net syn or anti addition.

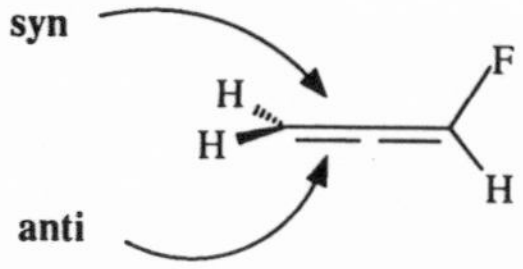

In recent years one has seen an increasing interest in what has come to be known as the π-*facial diastereoselectivity* of additions to unsymmetrically substituted π systems, particularly with regard to the effect of diastereotopically disposed *allylic* substituents.[96–99] Indeed, face selectivity in additions to trigonal carbon systems of all types is a truly fundamental question at the core of virtually all stereogenesis. It is therefore also of fundamental significance in organic synthesis as well as being of great portent to physical organic chemists. In fluoroallene, one has what is perhaps the perfect vehicle to probe the nature of π-facial selectivity in pericyclic reactions. First, in MFA the allylic fluorine substituent is perfectly aligned for either syn or anti electronic interaction. Second, the fluorine substituent of MFA, unlike any other substituent (other than deuterium, of course), is unlikely to exert a *steric* influence on the reactions being investigated. The only other systems that can compare in potential utility for probing π-facial selectivity are the 5-substituted adamantane derivatives being utilized by le Noble and his co-workers.[97,99] Indeed, le Noble has found that the fluorine substituent is also the most advantageous for these studies.

Diels–Alder Reactions

As indicated earlier, the Diels–Alder reactions of MFA are just as regiospecific with respect to selective reaction with its C_2–C_3 π bond as are the Diels–Alder reactions of DFA. With regard to *stereo*selectivity, our initial examination of MFA's reaction with cyclopentadiene did not indicate a significant preference for either syn or anti addition, but when we examined MFA's additions to butadiene and furan, a much more obvious syn selectivity was

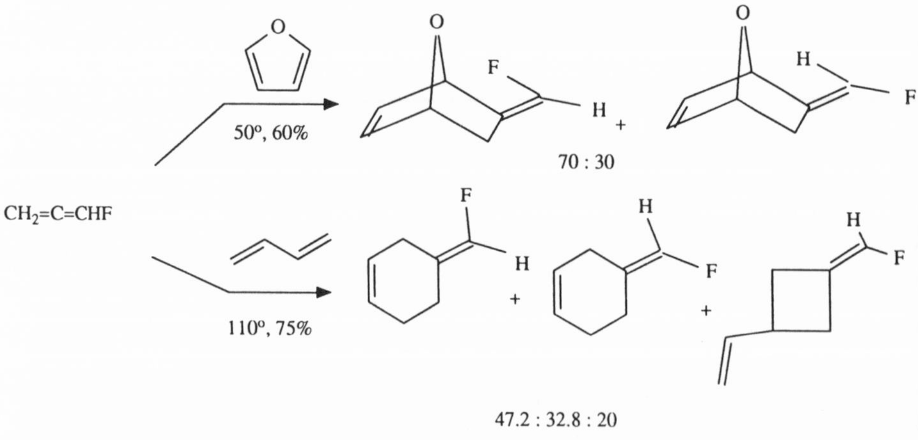

observed. Although these observed preferences for syn addition could not be called dramatic by any means, the transition state energy differences that would give rise to them being less than 1 kcal/mol, they were unprecedented at that time for Diels–Alder reactions. The possible sources of such preferences will be discussed later.

It should be noted that in MFA's reaction with 1,3-butadiene a minor [2 + 2] adduct was also observed to be formed in competition with the Diels–Alder reaction. Note that the regiochemistry of this adduct with respect to the MFA is *opposite* to that which was observed in the analogous reaction of DFA; that is, the fluorine-substituted double bond is in this case *not* preferentially incorporated into the ring. On the other hand the rationale that we used to explain the preferential incorporation of the CF_2 end of DFA into the ring of its [2 + 2] adduct is not obviated by the MFA result. On the contrary, the thermodynamic rationale that we used remains valid in that, while a CF_2 group has a definite strong thermodynamic preference for sp^3 over sp^2 hybridization, quite the opposite is true for the CHF group, as was discussed in Section 3.4.[47] Therefore, it would be predicted that in the [2 + 2] cycloadditions of MFA the generally preferred overall process should be reaction with its C_2–C_3 double bond.

[2 + 2] Cycloadditions

As predicted previously, regiochemically the [2 + 2] cycloadditions of MFA were observed, in contrast to those of DFA, to reflect a preference for the fluorine-substituted carbon to end up vinylic rather than in the ring.[56] Two

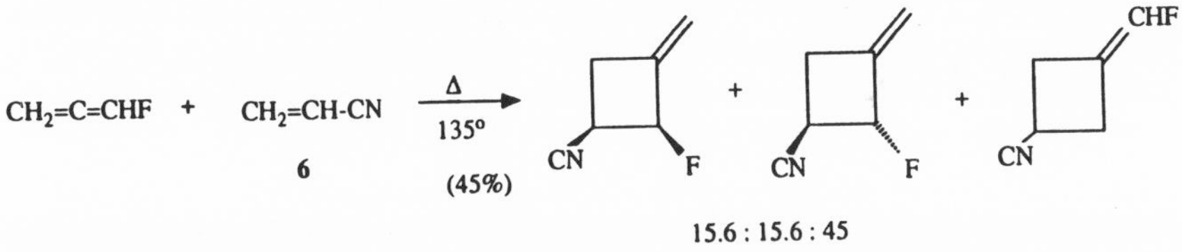

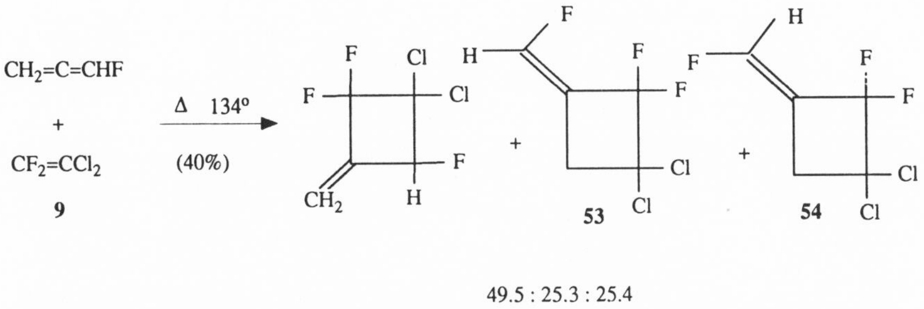

further examples, the reactions of MFA with acrylonitrile and with 1,1-dichloro-1,1-difluoroethylene, demonstrate that, as in the case of DFA, these reactions are not regiospecific but merely reflect the relative thermodynamic stability of the adducts.

In the reaction of MFA with 1,1-dichloro-2,2-difluoroethylene, it can be seen, from the ratios of products **53** and **54**, that there is virtually no preference observed for syn or anti addition in this [2 + 2] cycloaddition. Understanding that the factors which would determine π-facial selectivity for a multistep, diradical mechanism would be expected to be very different from those that would determine such in a concerted, pericyclic process, let us examine the results within the context of the probable mechanism. First, as indicated in our discussion of the mechanism of

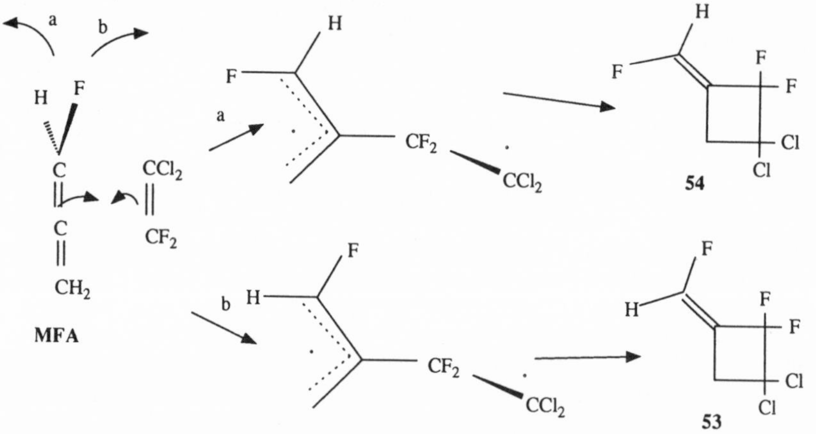

DFA's [2+2] reactions, initial bond formation should be to C_2 of the C_1–C_2 π bond, with the net stereochemical outcome of this reaction being determined by whether the fluorine substituent, in rotating into allylic conjugation, prefers to rotate toward or away from the attacking reagent. The results indicate that, for a substituent of such small steric demand as fluorine, no apparent rotational preference is observed. This is in marked contrast to camparable studies of monoalkyl allenes, wherein a definite preference for net anti addition has been reported and a steric rationale proposed.[38]

Thus, in a nonpericylic cycloaddition of MFA, wherein diradical intermediates are involved, no syn selectivity is observed. The observation of such selectivity in Diels–Alder and, as we will see, 1,3-dipolar cycloadditions is a strong indication that such selectivity derives from the pericyclic nature of the mechanism.

As for the *regio*selectivity observed in MFA's [2 + 2] reactions, as in the analogous DFA reactions, the relative ratio of adducts seems to reflect both the relative thermodynamic stability of the products and the relative stability of the expected diradical intermediates, with the expected more stable cyano-stabilized diradical leading to greater regioselectivity in its product formation.

1,3-Dipolar Cycloadditions

Cycloadditions of 1,3-dipoles being generally more orientationally selective than Diels–Alder reactions, we had hopes that greater syn/anti selectivity might be observed for these reactions as well. We were not to be disappointed.

Indeed, as we predicted earlier, MFA proved to be very reactive with nitrones, diazo compounds and nitrile oxides, although of course not as reactive as was DFA. MFA's reaction with *N-tert*-butylnitrone, for example, was complete after

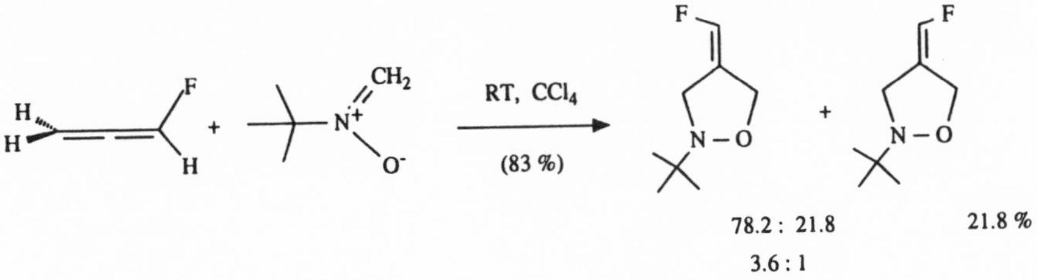

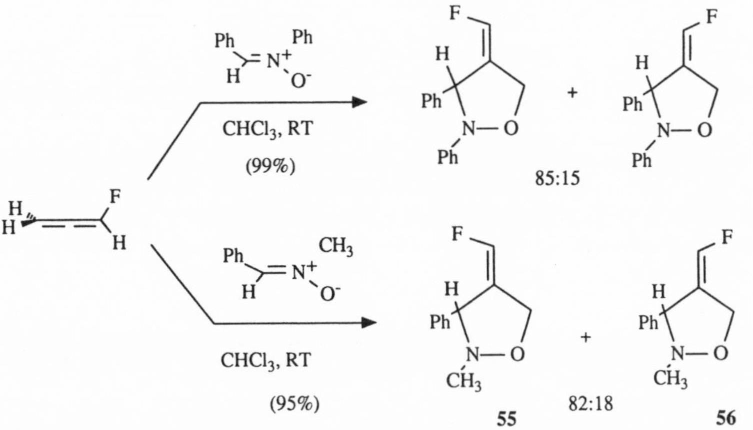

78.2 : 21.8   21.8 %

3.6 : 1

5 days at room temperature whereas the analogous reaction with DFA required only 16 h. To our satisfaction we observed that this reaction occurred with much greater syn selectivity (78 : 22) than had the Diels–Alder reaction.[63] Other nitrone cycloadditions occurred with even greater selectivity.[100,101]

The observed remarkable preference for syn addition of nitrones to MFA is in marked contrast to similar additions of nitrones to methoxy- and phenoxyallene, wherein anti products were formed to large excess.[102] Such results were explained in terms of steric effects. Since a fluorine substituent would be expected to exert little if any steric effect, our observed syn preference must derive from other, more subtle influences. The results may be related to the observation of preferential syn addition of 1,3-dipoles to *cis*-3,4-

Table 6.   Activation Parameters for the Reaction of MFA with
N-Methyl-C-phenylnitrone

Solvent	Reaction	$\Delta G^{\neq}$ (kcal/mol)	$\Delta H^{\neq}$	$\Delta S^{\neq}$ (eu)
Benzene-d_6	Overall	23.8	14.3	−32.0
	55	24.0	14.3	−32.4
	56	24.8	14.3	−35.2
CDCl$_3$	Overall	24.4	15.8	−28.8
	55	24.5	15.7	−29.5
	56	25.4	16.2	−30.8
Acetone-d_6	Overall	24.4	14.6	−32.7
	55	24.7	14.8	−33.3
	56	24.9	14.5	−35.3

dichlorocyclobutene,[103,104] although it should be noted that its Diels–Alder reactions exhibited no similar syn selectivity.[105]

Even though an 80 : 20 ratio is significant in terms of relative quantities, it nevertheless does not reflect a large difference in free energy of activation for the competitive processes. Activation parameters for the reaction of MFA with *N*-methyl-*C*-phenylnitrone were determined in three solvents and the results are given in Table 6. As one can see, a difference of less than 1 kcal/mol is required to explain all of the results in this work. One must then be very careful in attributing a unique factor to the observed effect. We will, however, with some trepidation, do our best to provide an explanation later in this chapter.

In the course of our kinetic studies, we investigated the effect of solvent polarity and solvent "character" upon rates and product ratios in the hope of being able to perceive a clear-cut distinction among those factors that might

Table 7.   Rates and Product Ratios in Various Solvents for the Reaction of
MFA with *N*-Methyl-*C*-phenylnitrone at 24°C

Solvent	E_T (kcal/mol)	k ($\times 10^6$)	k_{rel}	Product Ratio 55 : 56
Benzene-d_6	34.5	28.1	16	80 : 20
Dioxane-d_8	36.0	19.5	11	75 : 25
CDCl$_3$	39.1	9.58	5.4	82 : 18
CD$_2$Cl$_2$	41.1	9.48	5.3	76 : 24
Acetone-d_6	42.2	9.57	5.4	60 : 40
Me$_2$SO-d_6	45.0	9.22	5.2	53 : 47
CD$_3$CN	46.0	8.77	4.9	61 : 39
CD$_3$OD	55.5	1.78	1.0	68 : 32

have given rise to the syn selectivity. Table 7 gives the results of this solvent study.

In terms of rate effects, it would appear that a general decrease in overall rate as one progresses toward more polar solvents is observed. This is consistent with Huisgen's earlier studies of solvent effects on nitrone cycloadditions.[106] In fact, where common solvents were used in Huisgen's and our work a very similar trend was observed. Our activation energies were also consistent with those reported by Huisgen. Such comparisons indicate that the fluoroallene–nitrone cycloadditions should not be considered mechanistically different from other nitrone–alkene cycloadditions.

Looking at the effect of solvent polarity on syn/anti ratio, one can see that there is a definite trend toward less selectivity as one goes toward more polar solvents. However, there seems to be unusual behavior exhibited in three solvents—acetone, dimethylsulfoxide (DMSO), and acetonitrile—wherein unexpectedly large relative rates for formation of the anti isomer are observed. Interestingly these three solvents are distinguishable in character from the remaining solvents in that they have polar π bonds, which are potentially capable of giving rise to a mechanistically specific "complexation/solvation" of the nitrone 1,3-dipolar species. What effect this might have upon reactivity, specifically to form the minor isomer, is an interesting question that cannot be answered at this time.

Pertinent to this question is a comparison of the cycloaddition activation parameters in the solvents benzene and acetone. In benzene, it would appear that the syn/anti product ratio is due not to an enthalpic difference but to a difference in entropy of activation. On the other hand, in going to acetone as a solvent, the entropic preference for syn largely remains, but now the anti isomer becomes slightly favored enthalpically, hence the observed relative enhancement of anti reactivity. The differences in reactivities of nitrones in noninteractive (i.e., CH_2Cl_2, dioxane, or benzene) and interactive solvents (acetone, DMSO, or acetonitrile) might be due to the dipole acting as its own "solvator" in the noninteracting solvents, and thus perhaps existing as clusters that could have significantly different reactivity characteristics. Nevertheless, the experimental results obtained thus far do not allow unambiguous conclusions to be reached.

The cycloaddition chemistry of MFA with diazoalkanes bears characteristics of DFA's complex behavior with diazoalkanes. As in its reaction with DFA,

88:12

Table 8.   Regio- and Stereochemistry of Diazoalkane Cycloadditions to MFA

Diazoalkane	Relative Yields of Adducts				Regisomeric Ratio	Total Yield (%)
	57	58	59	60		
$CD_2=N_2$	88	12	—	—	∞	54
31a	37.8	23.3	4.2	34.7	10.5	92
31b	11.2	20.6	8.8	59.3	0.46	98
31c	2.7	6.9	8.9	81.5	0.11	94

diazomethane itself gives only a single regioisomeric adduct in reacting with MFA. Indeed, if one uses deuterium-labeled diazomethane in the reaction, a strong syn selectivity is observed, a result consistent with the abovementioned nitrone chemistry.

As discussed earlier in the chapter, results from the cycloaddition reactions of DFA with diazo compounds indicated that substituents on the diazo com-

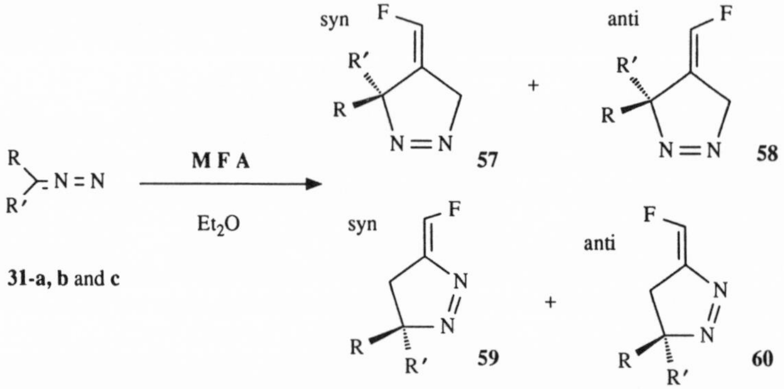

pounds gave rise to increasing proportions of the more sterically favored regioisomeric adducts. Cycloadditions of these same substituted diazo compounds with MFA led to even more insightful results: Not only do the sterically favored regioisomers become predominant as the substituents become bulkier,

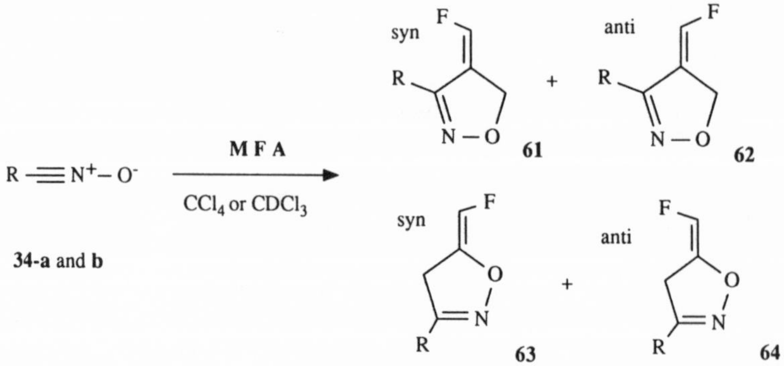

Table 9.   Regio- and Stereochemistry of Nitrile Oxide
Cycloadditions to MFA

Nitrile Oxide	Relative Yields of Products				Regioisomeric Ratio	Total Yield (%)
	61	62	63	64		
34a	4.0	35.8	16.7	43.5	0.65	88
34b	4	85	4	7	8.1	93

but the anti stereoisomers **58** and **60** also are seen to predominate. Table 8 summarizes the results.

In MFA's reactions with phenyl and mesityl nitrile oxides, both regioisomers were formed for both nitrile oxides, but the most interesting observation was that anti addition predominated for the formation of all of the regioisomeric adducts. Table 9 gives the results.

Rationale for Regiochemistry and Stereochemistry

The same rationale that was effective in explaining the regiochemistry of DFA's 1,3-dipolar cycloadditions may be applied to the 1,3-dipolar cycloadditions of MFA. In essence, the results are consistent with an intrinsic preference for regiospecificity, which is due to electrostatic interactions being capable of being overwhelmed by steric effects in the case of the diazoalkane reactions.

However, one must address the even more delicate question of what factors affect the determination of syn/anti selectivity in the 1,3-dipolar cycloadditions of MFA. In answering this question one must consider those same factors that were previously discussed as being important in the determination of the regiochemistry, that is, molecular orbital effects, steric effects, and electrostatic interactions.

Regarding molecular orbital effects, for much the same reasons as discussed with regard to regiochemical determination, it does not appear that frontier molecular orbital interactions can reasonably explain the syn/anti selectivity observed. The relative lack of significant selectivity in the Diels–Alder additions of MFA to hydrocarbon dienes was a critical observation in reaching this conclusion.

Steric effects cannot, of course, be involved in those reactions that give rise to syn selectivity, since such results are contrasteric. Even in the cases that give rise to favored anti addition, it is unlikely that the determining factor is a steric one, as will be discussed shortly.

With regard to the importance of electrostatic interactions, Figures 1 and 2 provide approximate pictoral representations of the calculated electrostatic interactions that we calculated to be present for the competing syn and anti transition states for nitrone addition. As can be seen, in both orientations, the

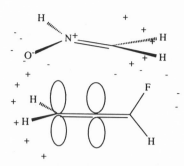

Figure 1. Electrostatic interactions for the syn transition state of the MFA–nitrone cycloaddition.

O-C_3 interactions is favorable, but the attractive F-CH_2 interaction in the syn orientation is clearly more favorable than the repulsive C_1-CH_2 interaction in the anti orientation. Such calculations indicate that electrostatic interactions are potentially an important factor, not only for the regioselectivity that is observed in 1,3-dipolar cycloadditions of DFA and MFA, but also for the stereoselectivity of MFA cycloadditions.

Certainly the syn selectivity observed in all of MFA's nitrone cycloadditions is consistent with electrostatic considerations as depicted in Figures 1 and 2. Moreover, on the basis of the MFA/diazomethane reaction there would seem to be an intrinsic favoring of syn addition for diazo compounds, likely owing to similar electrostatic factors. Even dimethyldiazomethane showed some syn preference in its formation of the "normal" regioisomer **57**. The switch to preferential anti addition in forming this regioisomer, for diphenyl diazomethane and for diazofluorene, could be due to the intervention of a steric

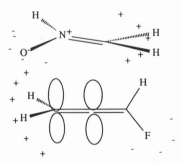

Figure 2. Electrostatic interactions for the anti transition state of the MFA–nitrone cycloaddition.

fluorine effect, but it is more likely due to electrostatic repulsion of the aromatic π electrons and the fluorine substituent in the syn transition state, much as we suggested to rationalize the reversal of regiochemistry in nitrile oxide cycloadditions to DFA. As for the "alternate" adducts **59** and **60**, with the nitrogens of the diazo compounds directed toward the fluorinated end of the MFA in this reaction it is likely that electrostatic repulsion of the nitrogen with the fluorine substituent in the syn transition state leads to the anti preference, which is generally observed in the formation of this regioisomer. The same argument holds for the preference of anti addition for the formation of regioisomer **64** in the nitrile oxide additions to MFA. As for the anti preference seen in formation of the "normal" regioisomer **62** in these reactions, one must again invoke electrostatic repulsion of the aromatic π electrons in the linear nitrile oxides as the determining factor for this observed stereoselectivity. Thus electrostatic interactions can be used to explain very effectively both the syn and the anti π-facial selectivities that are observed in the 1,3-dipolar cycloadditions of MFA.

Before we close the issue, however, we should consider another factor that has recently come into favor in explaining π-faciality. This factor would fall into the category of what are known as "secondary orbital effects," and it has been referred to by Cieplak,[96] le Noble,[97,99] and others[98] as the "hyperconjugation factor." Simply stated this factor predicts that "both nucleophiles and electrophiles [should] approach trigonal carbon from the direction antiparallel to the electrorichest [allylic] single bond."[99] This prediction was predicated upon the principle that an antiperiplanar σ bond should be able to delocalize into the newly developing σ* orbital (created in our case upon bond formation at C_2 of the allene).

Le Noble, moreover, found that Diels–Alder reactions adhered to this prediction,[97] which when applied to cycloadditions of MFA would certainly predict syn π-facial selectivity since an antiparallel C–H σ bond should definitely be better hyperconjugating than a C–F σ bond.

It thus appears that the syn π-facial diastereoselectivity that is observed in our 1,3-dipolar cycloadditions to MFA is, at least at the first level, consistent with Cieplak's and le Noble's hyperconjugation hypothesis. However, the switches from syn to anti preference that we have observed in the diazoalkane and nitrile oxide cycloadditions would not appear to be as easily reconciled by the hyperconjugation hypothesis as by electrostatic interaction arguments.

It is always a problem in designing an experiment to give a unique answer. The use of MFA in probing π-facial selectivities surmounts, for the most part, steric ambiguities of the type that are present with virtually all other substituent probes. However, in the case at hand, because a fluorine substituent will undoubtedly exert a powerful electrostatic effect, it is virtually impossible to differentiate experimentally between such an electrostatic explanation and the unrelated hyperconjugation possibility.

Thus we come full circle and return to the beauty of a deuterium substituent probe of mechanism. A deuterium substituent will, of course, not exert an electrostatic effect, nor a significant steric effect. There would also not be any molecular orbital differences, either frontier or secondary, between a hydrogen and a deuterium substituent. However, it has been amply demonstrated that a

deuterium substituent gives rise to a substantial hyperconjugative effect relative to hydrogen. In a future study, then, we will be examining the syn/anti selectivity for Diels–Alder and 1,3-dipolar cycloadditions to the C_2–C_3 double bond of monodeuterioallene. If the hyperconjugative explanation for π-facial selectivity is indeed legitimate, it should be apparent through a preferential formation of adduct 65.

4. CONCLUSIONS

In this chapter, we have attempted to demonstrate, through the specific example of our twenty-three years of research in the area of allene cycloadditions, the evolution of a typical mechanistic project from conception to fruition: the thought processes used in designing and carrying out the project, the extreme importance of persevering and maintaining an open mind throughout, and the inevitable and occasionally drastic turns that an even carefully planned project can take—in our case from use of deuterium isotope effects to fluorine substituent effects, from molecular orbital effects to electrostatic interactions and secondary molecular orbital effects.

ACKNOWLEDGMENTS

The research described in this chapter has been generously supported by the National Science Foundation and by the Petroleum Research Fund, which is administered by the American Chemical Society, and this support is gratefully acknowledged by the author.

The author also wishes to acknowledge all of his co-workers who conscientiously and capably carried out the described research.

REFERENCES

1. Hoffmann, R.; Woodward, R. B. *J. Am. Chem. Soc.* **1965**, *87*, 2046.
2. Hoffmann, R.; Woodward, R. B. *Acc. Chem. Res.* **1968**, *1*, 17.
3. Montgomery, L. K.; Schueller, K.; Bartlett, P. D. *J. Am. Chem. Soc.* **1964**, *86*, 622.
4. Bartlett, P. D. *Science* **1968**, *159*, 833.
5. Roedig, A. *Angew. Chem., Int. Ed. Engl.* **1969**, *8*, 150.
6. Kiefer, E. F.; Okamura, M. Y. *J. Am. Chem. Soc.* **1968**, *90*, 4187.
7. Moore, W. R.; Bach, R. D.; Ozretich, T. M. *J. Am. Chem. Soc.* **1969**, *91*, 5918.
8. Baldwin, J. E.; Roy, U. V. *J. Chem. Soc. D* **1969**, 1225.
9. Roedig, A.; Defzer, N. *Justus Liebigs Ann. Chem.* **1967**, *710*, 1.
10. Gajewski, J. J.; Black, W. A. *Tetrahedron Lett.* **1970**, 899.
11. Jacobs, T. L.; McClenon, J. R.; Muscio, O. J., Jr. *J. Am. Chem. Soc.* **1969**, *91*, 6038.
12. Dehmlow, E. V. *Tetrahedron Lett.* **1969**, 4283.
13. Huisgen, R.; Otto, P. *J. Am. Chem. Soc.* **1968**, *90*, 5342.
14. Brady, W. T.; Roe, R., Jr. *J. Am. Chem. Soc.* **1970**, *92*, 4618.
15. Frey, H. M.; Isaacs, N. S. *J. Chem. Soc. B* **1970**, 830.
16. Baldwin, J. E.; Kapecki, J. A. *J. Am. Chem. Soc.* **1970**, *92*, 4868.
17. Montaigne, R.; Ghosez, L. *Angew. Chem., Int. Ed. Engl.* **1968**, *7*, 221.
18. Kistiakowsky, G. B.; Ruhoff, J. R.; Smith, H. A.; Vaughan, W. E. *J. Am. Chem. Soc.* **1936**, *58*, 146.
19. Pledger, H. J. Org. Chem. **1960**, *25*, 278.
20. Carboni, R. A.; Lindsey, R. V. *J. Am. Chem. Soc.* **1959**, *81*, 4342.
21. Domelsmith, L. N.; Houk, K. N.; Piedrahita, C. A.; Dolbier, W. R. Jr. *J. Am. Chem. Soc.* **1978**, *100*, 6908.
22. Cripps, H. N.; Williams, J. K.; Sharkey, W. H. *J. Am. Chem. Soc.* **1959**, *81*, 2723.
23. Houk, K. N. *J. Am. Chem. Soc.* **1973**, *95*, 4092.
24. Houk, K. N. *Acc. Chem. Res.* **1975**, *8*, 361.
25. Van Sickle, D. E.; Rodin, J. O. *J. Am. Chem. Soc.* **1964**, *86*, 3091.
26. Selzer, S. *J. Am. Chem. Soc.* **1965**, *87*, 1534.
27. Katz, T. J.; Dessau, R. *J. Am. Chem. Soc.* **1963**, *85*, 2172.
28. Bayne, W. F.; Snyder, E. I. *Tetrahedron Lett.* **1970**, 2263.
29. Brown, P.; Cookson R. C. *Tetrahedron* **1965**, *21*, 1993.
30. Pryor, W.; Henderson, L. W. *Int. J. Chem. Kinet.* **1972**, *4*, 325.
31. Crawford, R. J.; Cameron, D. M. *J. Am. Chem. Soc.* **1966**, *88*, 2589.
32. Dolbier, W. R., Jr.; Dai, S. H. *J. Am. Chem. Soc.* **1968**, *90*, 5028.
33. Dolbier, W. R., Jr.; Dai, S. H. *J. Am. Chem. Soc.* **1970**, *92*, 1774.
34. Dolbier, W. R., Jr.; Dai, S. H. *Tetrahedron Lett.* **1970**, 4645.
35. Dai, S. H.; Dolbier, W. R., Jr. *J. Am. Chem. Soc.* **1972**, *94*, 3946.
36. Pasto, D. J. *J. Am. Chem. Soc.* **1979**, *101*, 37.
37. Pasto, D. J.; Heid, P. F.; Warren, S. E. *J. Am. Chem. Soc.* **1982**, *104*, 3676.
38. Pasto, D. J.; Warren, S. E. *J. Am. Chem. Soc.* **1982**, *104*, 3670.
39. Pasto, D. J.; Yang, S. H. *J. Org. Chem.* **1986**, *51*, 1676.
40. Schlag, E. W.; Peatman, W. B. *J. Am. Chem. Soc.* **1964**, *86*, 1676.
41. Chambers, R. D. *Fluorine in Organic Chemistry*; Wiley: New York, 1973.
42. Knoth, W. H.; Coffman, D. D. *J. Am. Chem. Soc.* **1960**, *82*, 3872.
43. Smart, B. E. In *The Chemistry of Functional Groups*; Patai, S.; Rappoport, Z., Eds.; Wiley: New York, 1983; Suppl. D, Chapter 14, pp 603–605.

44. Smart, B. E. In *Molecular Structure and Energetics*; Liebman, J.; Greenberg, A., Eds.; VCH Publishers: New York, 1986; Vol. 3, Chapter 14, pp 141–191.
45. Patrick, C. R. *Adv. Fluorine Chem.* **1961**, *2*, 1.
46. Hirsch, J. A. *Top. Sterochem.* **1967**, *1*, 199.
47. Dolbier, W. R., Jr.; Medinger, K. S.; Greenberg, A.; Liebman, J. F. *Tetrahedron* **1982**, *38*, 2415.
48. Houk, K. N. *J. Am. Chem. Soc.* **1973**, *95*, 4092.
49. Houk, K. N. *Acc. Chem. Res.* **1975**, *8*, 361.
50. Dolbier, W. R., Jr.; Piedrahita, C. A.; Houk, K. N.; Strozier, R. W.; Gandour, R. W. *Tetrahedron Lett.* **1978**, 2231.
51. Domelsmith, L. N.; Houk, K. N.; Piedrahita, C. A.; Dolbier, W. R., Jr. *J. Am. Chem. Soc.* **1978**, *100*, 6908.
52. Dixon, D. A.; Smart, B. E. *J. Phys. Chem.* **1989**, *93*, 7772.
53. Brundle, C. R.; Robin, M. B.; Kuebler, N. A.; Basch, H. *J. Am. Chem. Soc.* **1972**, *94*, 1451.
54. Brundle, C. R.; Robin, M. B.; Kuebler, W. A. *J. Am. Chem. Soc.* **1972**, *94*, 1466.
55. Bartlett, P. D.; Wingrove, A. S.; Owyang, R. *J. Am. Chem. Soc.* **1968**, *90*, 6067.
56. Dolbier, W. R., Jr.; Burkholder, C. R. *J. Org. Chem.* **1984**, *49*, 2381.
57. Burkholder, C. R. Ph.D. Dissertation, University of Florida, 1984.
58. Houb, K. N.; Sims, J.; Duke, R. E., Jr.; Strozier, R. W.; George, J. K. *J. Am. Chem. Soc.* **1973**, *95*, 7289.
59. Sustmann, R. *Tetrahedron Lett.* **1971**, 2717.
60. Huisgen, R.; Grashey, R.; Sauer, J. In *The Chemistry of Alkenes*; Patai, S., Ed.; Interscience: London, 1964; p 739.
61. Huisgen, R. *Angew. Chem., Int. Ed. Engl.* **1963**, *2*, 565, 633.
62. Dolbier, W. R., Jr.; Burkholder, C. R. *Israel J. Chem.* **1985**, *26*, 115.
63. Dolbier, W. R., Jr.; Seabury, M. J.; Burkholder, C. R.; Wicks, G. E.; Purvis, G. D., III. *Tetrahedron Lett.* **1990**, *46*, 7991.
64. Dolbier, W. R., Jr.; Burkholder, C. R.; Winchester, W.R. *J. Org. Chem.* **1984**, *49*, 1518.
65. Klopman, G. *J. Am. Chem. Soc.* **1968**, *90*, 223.
66. Salem, L. *J. Am. Chem. Soc.* **1968**, *90*, 543, 553.
67. Houk, K. N.; Sims, J.; Watts, C. R.; Luskus, L. J. *J. Am. Chem. Soc.* **1973**, *95*, 7301.
68. Kahn, S. D.; Pau, C. F.; Overman, L. E.; Hehre, W. J. *J. Am. Chem. Soc.* **1986**, *108*, 7381.
69. Kahn, S. D.; Hehre, W. J. *J. Am. Chem. Soc.* **1987**, *109*, 663.
70. Fisher, M. J.; Hehre, W. J.; Kahn, S. D.; Overman, L. E. *J. Am. Chem. Soc.* **1988**, *110*, 4625.
71. DeBenedetti, P. G.; DeMicheli, C.; Gandolfi, R.; Gariboldi, P.; Rostelli, A. *J. Org. Chem.* **1980**, *45*, 3646.
72. Firestone, R. A. *J. Org. Chem. Soc.* **1968**, *33*, 2285.
73. Firestone, R. A. *Tetrahedron* **1977**, *33*, 3009.
74. Piedrahita, C. A., Ph.D. Dissertation, University of Florida, 1978. 75. Epiotis, N. E. *J. Am. Chem. Soc.* **1972**, *94*, 1924, 1935.
76. Roberts, J. D.; Sharts, C. *Org. React. (N.Y.)* **1962**, 12, 1.
77. Heiba, E. I. *J. Org. Chem.* **1966**, *31*, 776.
78. Pasto, D. J.; Warren, S. E. *J. Org. Chem.* **1981**, *46*, 2842.
79. Pasto, D. J.; Warren, S. E.; Morrison, M. A. *J. Org. Chem.* **1981**, *46*, 2837.
80. Hopf, H. In *The Chemistry of Allenes*; Landor, S. D., Ed.; Wiley-Interscience: New York, 1982; Vol. 2, p 525.
81. Dolbier, W. R., Jr.; Seabury, M. J.; Daly, D.; Smart, B. E. *J. Org. Chem.* **1986**, *51*, 974.
82. Pasto, D. J.; Yang, S. H. *J. Am. Chem. Soc.* **1984**, *106*, 152.
83. Dolbier, W. R., Jr.; Wicks, G. E. *J. Am. Chem. Soc.* **1985**, *107*, 3626.
84. Creary, X. *J. Org. Chem.* **1980**, *45*, 280.

85. Dolbier, W. R., Jr.; Seabury, M. J. *Tetrahedron Lett.* **1987**, *28*, 1491.
86. Dolbier, W. R., Jr.; Burkholder, C. R. *J. Am. Chem. Soc.* **1984**, *106*, 2139.
87. Chang, M. H.; Crawford, R. J. *Can. J. Chem.* **1981**, *59*, 2556.
88. Crawford, R. J.; Chang, M. H. *Tetrahedron* **1982**, *38*, 837.
89. Dolbier, W. R., Jr.; Phanstiel, O., IV. *J. Am. Chem. Soc.* **1989**, *111*, 4907.
90. Hasselmann, D. *Tetrahedron Lett.* **1972**, 3465.
91. Hasselmann, D. *Angew. Chem., Int. Ed. Engl.* **1975**, *14*, 257.
92. Klot, I. M.; Rosenberg, R. M. *Chemical Thermodynamics*, 4th ed.; Cummings: Reading, MA, 1986; p 277.
93. Dolbier, W. R., Jr.; Seabury, M. J. *J. Am. Chem. Soc.* **1987**, *109*, 4393.
94. Dolbier, W. R., Jr.; Weaver, S. L. *J. Org. Chem.* **1990**, *55*, 711.
95. Firestone, R. A.; Vitale, M. A. *J. Org. Chem.* **1981**, *46*, 2160.
96. Johnson. C. R.; Tait, B. D.; Cieplak, A. S. *J. Am. Chem. Soc.* **1987**, *109*, 5875.
97. Chung, W. S.; Turro, N. J.; Srivanstava, S.; Li, H.; le Noble, W. J. *J. Am. Chem. Soc.* **1988**, *110*, 7882.
98. Naperstkow, A. M.; Macaulay, J. B.; Newlands, M. J.; Fallis, A. G. *Tetrahedron Lett.* **1989**, *30*, 5077.
99. Srivastava, S.; le Noble, W. J. *J. Am. Chem. Soc.* **1987**, *109*, 5874.
100. Dolbier, W. R., Jr.; Wicks, G. E.; Burkholder, C. R.; Palenik, G. J.; Gawron, M. *J. Am. Chem. Soc.* **1985**, *107*, 7183.
101. Dolbier, W. R., Jr.; Wicks, G. E.; Burkholder, C. R. *J. Org. Chem.* **1987**, *52*, 2196.
102. Battioni, P.; VoQuang, F.; VoQuang, Y. *Bull. Soc. Chim. Fr.* **1978**, *401*, 415.
103. Gandolfi, R.; Ratti, M.; Toma, L. *Heterocycles* **1979**, *12*, 897.
104. Caramella, P.; Albini, F. M.; Vitali, D.; Rondan, N. G.; Wu, Y. D.; Schwartz, T. R.; Houk, K. N. *Tetrahedron Lett.* **1984**, *25*, 1875.
105. Warrener, R. N.; Johnson, R. P.; Jefford, C. W.; Ralph, D. A. *Tetrahedron Lett.* **1979**, *20*, 2939.
106. Huisgen, R.; Seidl, H.; Brunning, I. *Chem. Ber.* **1969**, *102*

INDEX

181

Advances in Molecular Vibrations and Collision Dynamics

Edited by **Joel M. Bowman**, *Department of Chemistry, Emory University*

Volume 1, 1991, 2 Volume Set
Set ISBN 1-55938-293-7 $157.00

Volume 1 - Part A, 1991, 360 pp. $78.50
ISBN 1-55938-294-5

CONTENTS: **An Introduction to the Dynamics of van der Waals Molecules,** *Jeremy M. Hutson, University of Durham.* **The Nature and Decay of Metastable Vibrations: Classical and Quantum Studies of van der Waals Molecules,** *Stephen K. Gray, Northern Illinois University.* **Optothermal Vibrational Spectroscopy of Molecular Complexes,** *R.E. Miller, University of North Carolina.* **High Resolution IR Laser Driven Vibrational Dynamics in Supersonic Jets: Weakly Bound Complexes and Intramolecular Energy Flow,** *Andrew McIlroy and David J. Nesbitt.* **Three Dimensional Quantum Scattering Studies of Transition State Resonances: Results for O + HCl OH + Cl,** *Hiroyasu Koizumi, Northwestern University and George C. Schatz, Argonne National Laboratory.* **Negative Ion Photodetachment as a Probe of the Transition State Region: The + HI Reaction,** *Daniel M. Neumark, University of California, Berkeley.* **Rovibrational Spectroscopy of Transition States,** *James J. Valentini, University of California, Irvine.* **Optimal Control of Molecular Motion: Making Molecules Dance,** *Herschel Rabitz and Shenghua Shi, Princeton University.* **Static Self Consistent Field Methods for Anharmonic Problems: An Update,** *Mark A. Ratner, Northwestern University, Robert B. Gerber, The Hebrew University and University of California, Irvine, Thomas R. Horn, Northwestern University and The Hebrew University, and Carl J. Williams, Northwestern University.* **Perturbative Studies of the Vibrations of Polyatomic Molecules Using Curvilinear Coordinates,** *Anne B. McCoy and Edwin L. Sibert III, University of Wisconsin-Madison.*

Volume 1 - Part B, 1991, 327 pp. $78.50
ISBN 1-55938-295-3

CONTENTS: **Preface. Classical Dynamics and the Nature of Highly Excited Vibrational Eigenstates,** *Michael J. Davis, Argonne National Laboratory, Craig C. Martens, University of California, Irvine, Robert G. Littlejohn and J.S. Pehling, University of California, Berkeley.* **Semiclassical Mechanisms of Bound and Unbound States of Atoms and Molecules,** *David Farrelly,*

University of California, Los Angeles. **Dissociation of Overtone-Excited Hydrogen Peroxide Near Threshold: A Quasiclassical Trajectory Study,** *Yuhua Guan, Brookhaven National Laboratory, Turgay Uzer, Brian D. Macdonald, Georgia Institute of Techhology, and Donald L. Thompson, Oklahoma State University.* **L2 Approaches to the Calculation of Resonances in Polyatomic Molecules,** *Bela Gazdy and Joel M. Bowman, Emory University.* **Analytic MBPT(2) Energy Derivaties: A Powerful Tool for the Interpretation and Prediction of Vibrational Sepctra for Unusual Molecules,** *Rodney J. Bartlett, John F. Stanton and John D. Watts, University of Florida, Gainesville.* **Spectro-A Program for the Derivation of Spectroscopic Constants from Provided Quartic Force Fields and Cubic Dipole Fields,** *Jeffrey F. Gaw, Monsanto Company, Andrew Willetts, William H. Green and Nicholas C. Handy, University Chemical Laboratory, England.* **Photoinitiated Reactions in Weakly-Bonded Complexes: Entrance Channel Specificity,** *Y. Chen, G. Hoffmann, S.K. Shin, D. Oh, S. Sharpe, Y.P. Zeng, R.A. Beaudet and C. Wittig, Universtiy of Southern California, Los Angeles.* **Photodissociation Dynamics of the Nitrosyl Halides: The Influence of Parent Vibrations,** *Charles Qian and Hanna Reisler, University of Southern California, Los Angeles.* **Gas-Phase Metal Ion Solvation: Spectroscopy and Simulation,** *James M. Lisy, University of Illinois at Urbana-Champaign.* **The Chemical and Physical Properties of Vibration-Rotation Eigenstates of** $H_2CO(SO)$ at 28,300 CM^{-1} *Hai-Lung Dai, University of Pennsylvania.*

JAI PRESS INC.

55 Old Post Road - No. 2
P.O. Box 1678
Greenwich, Connecticut 06836-1678
Tel: 203-661-7602